MULTIPLE PERIODIC VARIABLE STARS

ASTROPHYSICS AND SPACE SCIENCE LIBRARY

A SERIES OF BOOKS ON THE RECENT DEVELOPMENTS

OF SPACE SCIENCE AND OF GENERAL GEOPHYSICS AND ASTROPHYSICS

PUBLISHED IN CONNECTION WITH THE JOURNAL

SPACE SCIENCE REVIEWS

VOLUME 60

PROCEEDINGS

MULTIPLE PERIODIC VARIABLE STARS

PROCEEDINGS OF THE
INTERNATIONAL ASTRONOMICAL UNION COLLOQUIUM No. 29,
HELD AT BUDAPEST, HUNGARY 1–5 SEPTEMBER 1975

Edited by

WALTER S. FITCH
Steward Observatory University of Arizona

INVITED PAPERS

D. REIDEL PUBLISHING COMPANY
DORDRECHT–HOLLAND / BOSTON–U.S.A.

ISBN 90 277 0766 9

Joint edition published by
D. Reidel Publishing Company,
P.O. Box 17, Dordrecht, Holland and
Akadémiai Kiadó, Budapest, Hungary

Sold and distributed in the U.S.A., Canada, and Mexico
by D. Reidel Publishing Company, Inc.
Lincoln Building, 160 Old Derby Street, Hingham,
Mass. 02043, U.S.A.

Printed in Hungary

CONTENTS

PREFACE

The I.A.U. Colloquium No. 29 was held in Budapest, September 1–5, 1975. The subject of the colloquium was: "Multiple Periodic Variable Stars". The colloquium was organized by the Scientific Organizing Committee consisting of W. S. Fitch (Chairman), M. W. Feast, B. V. Kukarkin, P. Ledoux, J. Smak, R. S. Stobie, B. Szeidl, B. Warner and S. C. Wolff.

The local organization was placed in the hands of a Committee consisting of the staff of the Konkoly Observatory: B. Szeidl (Chairman), K. Barlai, M. Ill, S. Kanyó and L. Szabados.

The colloquium was attended by about 90 scientists representing Australia, Austria, Belgium, Bulgaria, Canada, Czechoslovakia, Denmark, France, F.R.G., G.D.R., Greece, Hungary, India, Ireland, Italy, Japan, The Netherlands, New Zealand, Poland, Roumania, South Africa, United Kingdom, U.S.A. and U.S.S.R.

Eight sessions were held, viz.

1. β Canis Majoris Stars. Chairman: W. Wenzel
2. Magnetic and Ap Variables. Chairman: S. C. Wolff
3. Miras and Red Variables. Chairman: P. R. Wood
4. Cepheids. Chairman: R. S. Stobie
5. RR Lyrae Stars. Chairman: B. Szeidl
6. RRs and δ Scuti Stars. Chairman: W. S. Fitch
7. White Dwarfs and Novae. Chairman: B. Warner
8. Close Binaries and X-Ray Sources. Chairman: E. H. Geyer

At the beginning of the first session a welcome was presented by Professor G. Marx on behalf of the Hungarian Academy of Sciences and the Eötvös Loránd Physical Society. Then B. Szeidl delivered a memorial lecture of László Detre who had initiated this colloquium.

In general, the sessions started with two reviews, one for theory and one for observation.

B. SZEIDL
Chairman of the Local Organizing Committee

LIST OF PARTICIPANTS

Australia

N. R. Lomb, Dept. of Astronomy, University of Sydney

Austria

A. Schnell, Universitäts-Sternwarte Wien

Belgium

M. Burger, Département d'Astrophysique, Université de Mons
E. Elst, Observatoire Royal de Belgique, Bruxelles
P. Smeyers, Astronomisch Instituut, Katholieke Universiteit Leuven

Bulgaria

G. Momchev, Yambol Observatory
N. Nikolov, Dept. of Astronomy, University of Sofia

Canada

G. A. Bakos, Dept. of Physics, University of Waterloo
K. Desikachary, Dept. of Astronomy, University of Western Ontario
J. L. Tassoul, Dept. de Physique, Université de Montreal
M. Tassoul, Dept. de Physique, Université de Montreal
A. F. Wehlau, Dept. of Astronomy, University of Western Ontario
W. H. Wehlau, Dept. of Astronomy, University of Western Ontario

Czechoslovakia

J. Grygar, Ondrejov Observatory
J. Tremko, Skalnate Pleso Observatory

Denmark

H. E. Jørgensen, University Observatory, Copenhagen
J. O. Petersen, University Observatory, Copenhagen

France

M. Auvergne, Observatoire le Mont-Gros, Université de Nice
A. Baglin, Observatoire le Mont-Gros, Université de Nice
M. Friedjung, Institut d'Astrophysique, Paris
J. M. Le Contel, Observatoire le Mont Gros, Université de Nice
M. Th. Martel, Observatoire de Lyon
J. P. Sareyan, Institut d'Astrophysique, Paris
J. C. Valtier, Observatoire de Nice

F.R.G.

E. H. GEYER, Observatorium Hoher List
W. P. GIEREN, Observatorium Hoher List
E. SCHOEFFEL, Remeis Sternwarte Bamberg
W. SEGGEWISS, Observatorium Hoher List
W. STROHMEIER, Remeis Sternwarte Bamberg
N. VOGT, Sternwarte Hamburg-Bergedorf

G.D.R.

G. HILDEBRANDT, Zentralinstitut für Astrophysik, Potsdam
W. SCHÖNEICH, Zentralinstitut für Astrophysik, Potsdam
W. WENZEL, Zentralinstitut für Astrophysik, Sternwarte Sonneberg

Greece

P. G. LASKARIDES, Laboratory of Astronomy, University of Athens

Hungary

K. BARLAI, Konkoly Observatory, Budapest
G. ERDŐS, Central Institute for Physics, Budapest
T. GOMBOSI, Central Institute for Physics, Budapest
E. ILLÉS, Konkoly Observatory, Budapest
S. KANYÓ, Konkoly Observatory, Budapest
L. PATKÓS, Konkoly Observatory, Budapest
L. SZABADOS, Konkoly Observatory, Budapest
B. SZEIDL, Konkoly Observatory, Budapest
GY. TÓTH, Gothard Observatory, Szombathely

India

H. S. MAHRA, Uttar Pradesh State Observatory

Ireland

P. A. WAYMAN, Dunsink Observatory, Dublin

Italy

C. BARTOLINI, Osservatorio Astronomico Universitario, Bologna
P. BROGLIA, Osservatorio Astronomico, Merate
C. CACCIARI, Osservatorio Astronomico Universitario, Bologna
P. CONCONI, Osservatorio Astronomico, Merate
M. FRACASSINI, Osservatorio Astronomico di Milano
A. GUARNIERI, Osservatorio Astronomico Universitario, Bologna
G. GUERRERO, Osservatorio Astronomico, Merate
L. E. PASINETTI, Osservatorio Astronomico, Merate

Japan

S. KATO, Dept, of Astronomy, University of Kyoto
M. TAKEUTI, Astronomical Institute, Tohoku University, Sendai

Netherlands

K. K. KWEE, Sterrewacht-Huygens Laboratorium, Leiden
J. H. WALRAVEN, Leiden Observatory, Southern Station
TH. WALRAVEN, Leiden Observatory, Southern Station

New Zealand

D. Bateson, Royal Astronomical Society of New Zealand
F. M. Bateson, Royal Astronomical Society of New Zealand

Poland

W. Dziembowski, Institute of Astronomy, Warsaw
S. Grudzinska, Astronomical Observatory, Torun
M. Jerzykiewicz, Astronomical Observatory, Wroclaw
T. Kozar, Astronomical Observatory, Wroclaw
M. Kozlowski, Institute of Astronomy, Warsaw
J. Krelowski, Astronomical Observatory, Torun
A. Opolski, Astronomical Observatory, Wroclaw
R. Tylenda, Astronomical Observatory, Torun

Rumania

A. R. Dinescu, Observatorul Astronomic, Bucuresti
I. Todoran, Observatorul Astronomic, Cluj-Napoca

South Africa

B. Warner, Dept. of Astronomy, University of Cape Town

United Kingdom

G. Th. Bath, Dept. of Astronomy, Oxford
E. Budding, Tokyo Astronomical Observatory, University of Tokyo
R. S. Stobie, Royal Observatory, Edinburgh
P. R. Wood, Astronomy Centre, University of Sussex

U.S.A.

A. N. Cox, Los Alamos Scientific Laboratory
J. P. Cox, JILA, University of Colorado
W. S. Fitch, Steward Observatory, University of Arizona
D. S. Hall, Dyer Observatory, Vanderbilt University
J. R. Lesh, Dept. of Physics and Astronomy, University of Denver
L. B. Lucy, Dept. of Astronomy, Columbia University
J. H. A. Mattei, AAVSO, Cambridge, Mass.
M. R. Molnar, Dept. of Physics and Astronomy, University of Toledo
H. M. Van Horn, Dept. of Physics and Astronomy, University of Rochester
J. R. Wolff, Institute for Astronomy, Honolulu
S. C. Wolff, Institute of Astronomy, University of Hawaii

U.S.S.R.

I. F. Alania, Abastumani Astrophysical Observatory
A. K. Alksnis, Radioastrophysical Observatory, Riga
M. S. Frolov, Sternberg Astronomical Institute, Moscow
E. Parsamian, Byurakan Astrophysical Observatory
I. B. Pustilnik, Astrophysical Observatory, Tartu
Y. Romanov, Astronomical Observatory, Odessa
V. P. Tsessevich, Astronomical Observatory, Odessa
U. Uus, Astrophysical Observatory, Tartu

Multiple Periodic Variable Stars
IAU Colloquium No.29, Budapest, 1975

IN MEMORY OF LÁSZLÓ DETRE (1906-1974)

B. Szeidl

Now we are just going to discuss the problems of multiple periodic variable stars. It is, however, very sad to know that the one who took the initiative in organizing this colloquium is no longer here among us. Professor Detre died last autumn when he was only at the age of 68.

Until the end of his life he was interested almost in every field of astronomy. Every day he devoted 5-6 hours to reading the periodicals and journals. I remember how enthusiastic he was when he had read about a new astronomical discovery.He was very well trained in mathematics and so it is not surprising that he was especially impressed by celestial mechanics.Poincaré's books were his favorite readings. But the variable stars had always been nearest to his heart. He had been engaged with variable star research during 40 years of his life.

Detre's career was not too eventful. In his youth he was already interested in sciences, especially in physics, mathematics and astronomy. Having finished his studies at the Friedrich Wilhelm Universität in Berlin he was appointed an assistant of the Budapest Observatory in 1929 and had been its director from 1943 until 1974.

During the years spent in Germany Dr. Detre made many friends. Even after many years he was still eager to return to Germany from time to time,and meet his old friends and see the places where he had lived. He liked to recall memories of the past, particularly the college-years and was fond of telling droll stories about the college-life in Berlin, famous professors and university-fellows who later on became well-known scientists.

Dr. Detre liked to talk much about his family-tree. Among his ancestors there were Germans as well as Armenians.But he in his heart was a real Hungarian. In the thirties when many Hungarian scientists left the country seeking better research possibilities and a more comfortable life, he remained here urged by a determination of raising the standard of astronomical research here in Hungary. He began working under very bad conditions. Having only a 6 in. photographic camera, already in 1933 he started the program which has been the main research field of the Konkoly Observatory during the past 40 years, i.e., the investigation of the light curve variations and period changes of a number of RR Lyrae stars. It was hoped that by scrutinizing both these problems a better insight could be obtained into the unknown mechanism of these stars. During the coming days we shall be able to see how these hopes have been fulfilled.

Dr. Detre was aware of the importance of the observations. If the sky was clear the telescopes at our observatory had to be in action both on weekdays and holidays. You can look up the records made at Christmas Eves or New Year's Eves and you will be convinced that not a single hour of clear nights was ever wasted. He used to work at night and it was always very pleasant to be in his company. He liked joking and talking. But if the sky was getting clear he suddenly changed. At such a time he was more like a slave-driver. Perhaps observing was the only matter he took very seriously.

In the 1950-s the observing conditions were getting worse and worse near Budapest. Under Professor Detre's directorship and direct guidance a new observing station was built in the Mátra mountains, on one of the highest points of Hungary. Now this new observatory is fairly well equipped. We have there a 20 in. Cassegrain-, a 24 in./36 in. Schmidt- and a 40 in. Ritchey-Chrétien telescope, all made by Zeiss-Jena.

Almost half a century ago, when Dr. Detre commenced work he had to overcome many difficulties. During these it was he who turned our unknown observatory into a well-equipped and better known institute. A year ago or so Professor Detre passed on to us the torch to carry. We hope that under the better conditions created by him we will be able to continue his work and be faithful to his heritage.

INVITED PAPERS

Multiple Periodic Variable Stars
IAU Colloquium No.29, Budapest, 1975

ON DOUBLE PERIODICITY IN THE LIGHT VARIATION OF VARIABLE STARS

V.P. Tsessevich

Odessa Astronomical Observatory, Odessa-270014, USSR

Double periodicity of light variations is shown in different ways in stars of different types. So the light curves of μ Cephei of RV Tauri-type giant stars are composed of more or less fast waves superposed on slow ones, such that between the duration of the short and long cycles there is a correlation.

Convincing evidences have been recently given as to the existence of superpositions of rapid fluctuations of eruptive character upon the slow ones; this makes light variations similar to solar activity manifestations.

Double periodicity of another type - interference of radius fluctuations occurring with similar periods which produces more or less long beat cycles - is inherent to the stars of β CMa-type, to dwarf cepheids and RR Lyrae-type stars.

Periods of RR Lyrae-type stars and of cepheids suffer slow secular changes. True, a view-point is common according to which the diagram of O-C residuals plotted against E represents a random phenomenon. One can hardly believe, however, that the O-C diagram of Z CVn or RZ Cep could be accounted for by the summation of cumulative errors. But variations of a period during "critical" moments can be fortuitous indeed and may result from abrupt stellar structure change.

Convinced we are in this due to the following items:

As shown in AP Her the O-C residual fluctuations are nearly of periodic character with the cycle about 800P. It is impossible to attribute these periodic fluctuations to fortuitous accumulation of errors. Neither is it possible to attribute them to evolutionary variation of stellar structure - they proceed

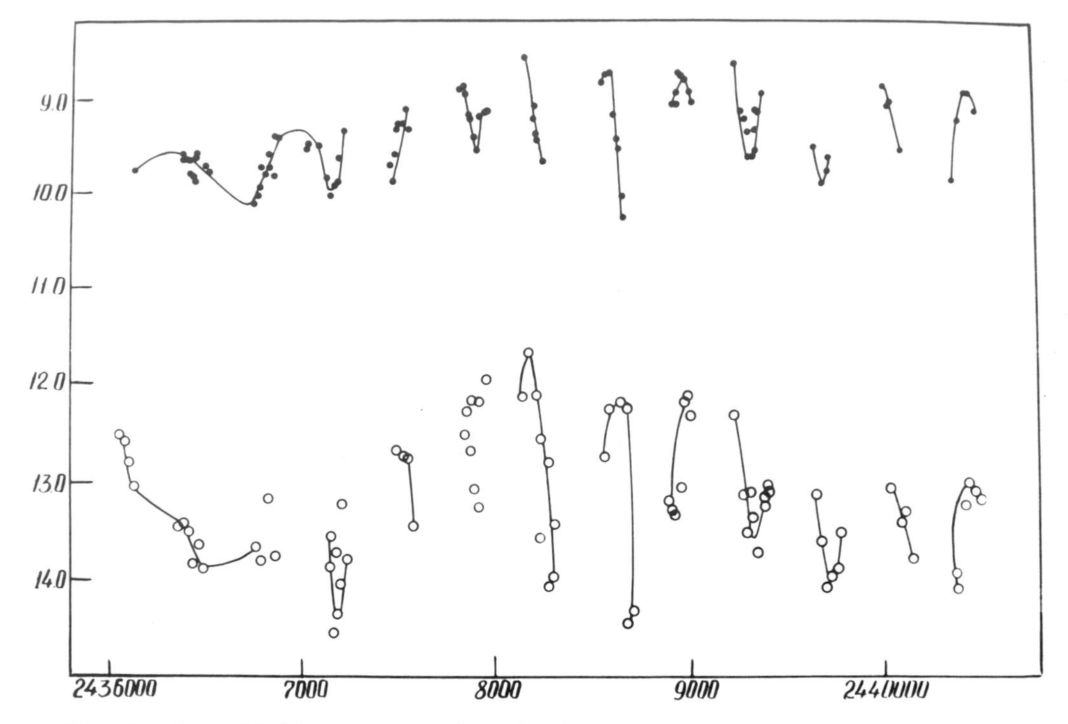

Fig.1. Two light curves of red giant EP Vul. Upper curve - photo-visual, lower-photographic one. On slow variations short ones are superposed.

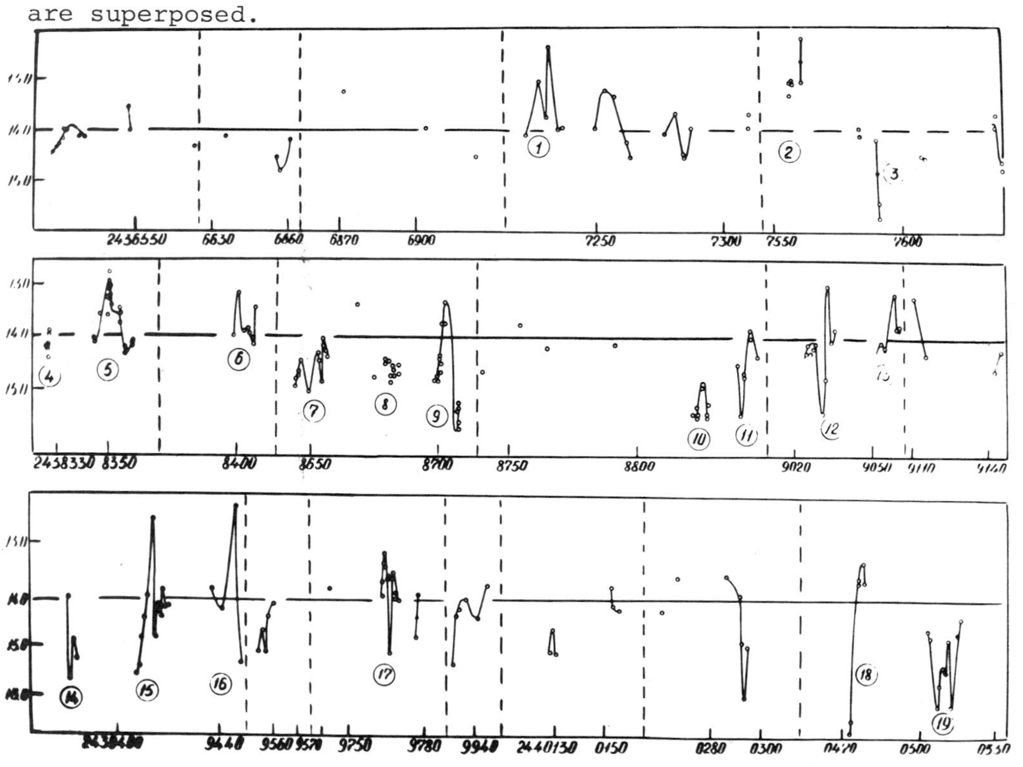

Fig.2. Fragments of light curve RW Aur-type star - AA Tau. There are flares superposed on slow variations of mean brightness.

too rapidly. The evolutionary track is also conspicuous in the diagram, with a slow secular increase accompanying the cyclic O-C behaviour.

Furthermore, the following phenomena have been observed repeatedly in RR Lyrae-type stars. After the second variation of the period the former value of period (or nearly the previous) was resumed. This suggests that a pulsating star shows two extreme states between which a real one is fluctuating. This conclusion is confirmed by the Blazhko-effect as well.

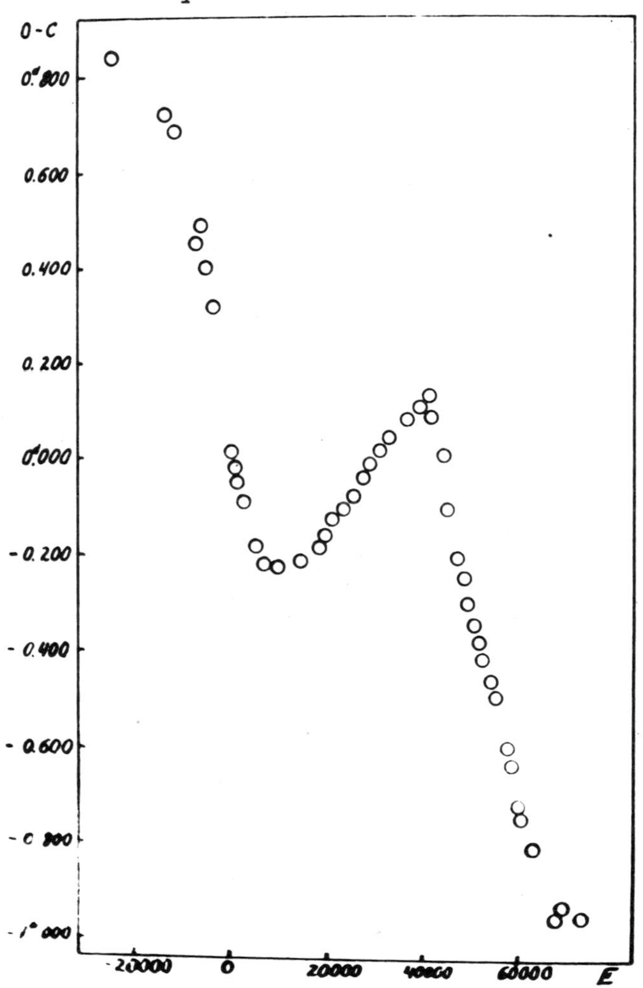

Fig.3. RZ Cep: residuals O-C plotted against number E.

The attempts of "summing" the Blazhko-effect of two interfering light fluctuations have failed. However, the star lumino-

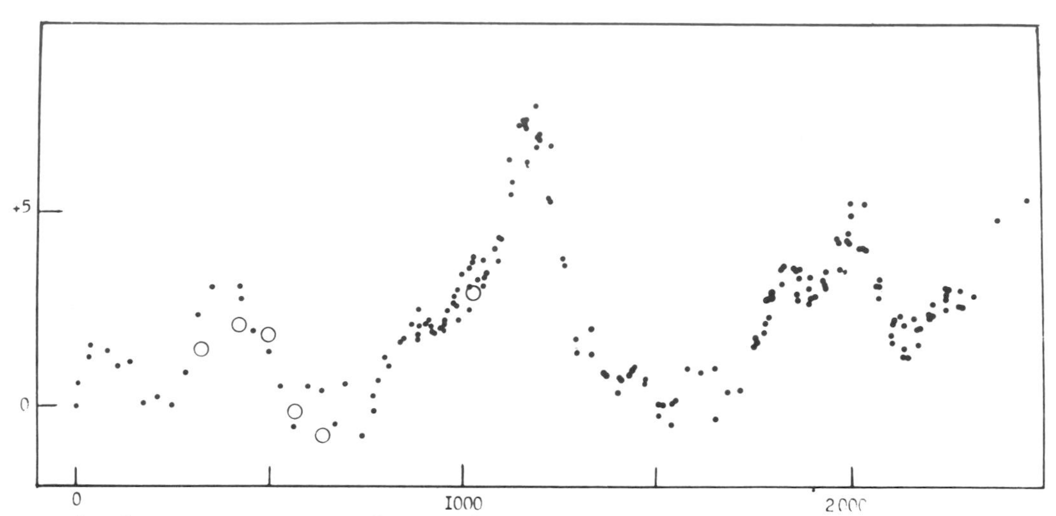

Fig.4. AP Her:star of CW-subtype. Periodical fluctuations of O-C residuals.

sity is not a primary value though composed of two primary values - radius and effective temperature. The most primary value in a pulsating star is its variable radius. The integration of curves of radial velocities has enabled us to find two oscillations of radius, the primary one with a large amplitude and the secondary one with a smaller amplitude. The addition of these two oscillations - primary and perturbing - brings about radius variation of a variable amplitude and the latter causes different type variations in luminosity.

Investigations of Blazhko-effect made it possible to divide it into some subtypes. Normal Blazhko-effect occurs with periodic variations both in the light curve and in the shift of the moment of maximum relative to the linear formula $M=M_0+PE$.

Anomalous Blazhko-effect involves amplitude variations without any shift of the moment of maximum or of variations in phase of maximum light without any variations in amplitude. Recently I have detected an anomalous Blazhko-effect in AR Ser. The Blazhko-effect period in this star varies irregularly from 90 to 120 days.

The O-C residuals are computed with respect to formula

$$\text{Max.hel. JD} = 2430472.469 + 0.57514215\ E.$$

The O-D residuals of moments of maximum amplitudes are computed with respect to formula

$$T_{Max,Amp.} = 243969.5 + 113.02\ E.$$

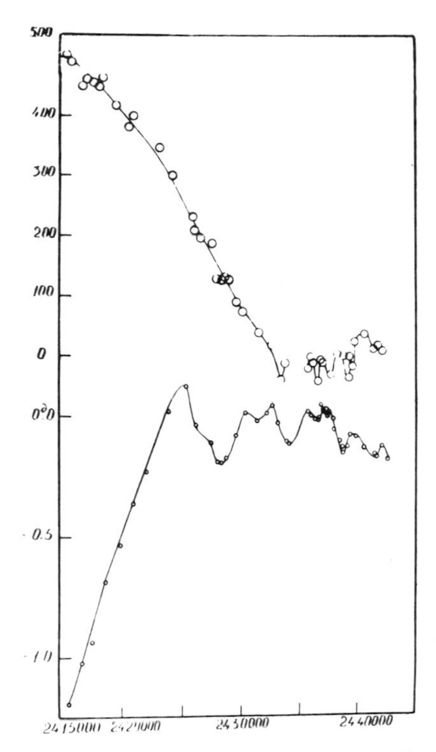

Fig.5. AR Ser: Residuals O-C and O-D plotted against JD (for details see text).

The light curve shape variations are distinctly shown in observations by Fitch, Wisniewski and Johnson.

It is stated that the period Π of Blazhko-effect changes essentially with the variation of period P of principal light variation. These variations are correlated which testifies to the fact of nonaccidental character of O-C graphs.

Thus Blazhko-effect indicates fortuitous reconstructions of internal stellar structure to occur, the cause of which is still to be found out. We have to assume, however, that variations in RR Lyrae-type stars are caused by fluctuations in the star's envelope, the suggestion being consistent with the present viewpoint of Zhevakin, Christy, et al.

These problems can be solved successfully if a sufficient number of observations are gained. Naturally, ample photoelectric observations of numerous objects are particularly urgent. At present, however, it seems next to impossible to realize all that. Therefore a considerable increase in visual observations

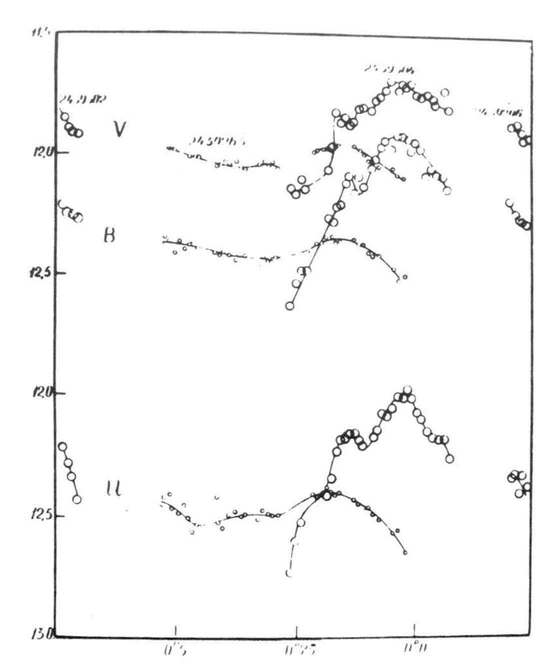

Fig.6. AR Ser: UBV light curves according to the observations by Fitch, Wisniewski, Johnson for two different phases of Blazhko-effect.

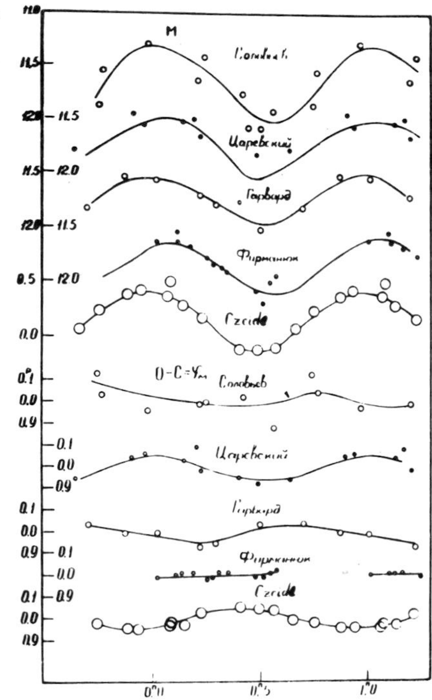

Fig.7. AR Ser: change of elements M and O-C during cycle of Blazhko-effect.

of many objects will be urgent too. At present, ample material is available due to the realization of IAU resolution suggested by L. Detre in 1957. The revised Catalogue is regularly published in Cracow as well as ephemerides' data calculated on its basis. This work has been carried out for more than 15 years. At first several observers used to send their observations to the yearly Catalogue revision which proved to be of great importance. Now the lot has fallen upon the Odessa Astronomical Observatory to participate in the activities, for many thousands of visual observations of over 200 objects are available here.

It is too extensive a problem to be solved by the colleagues of one observatory only. More and more groups of amateurs should be attracted for the purpose of collecting observations.

However, progressive collection of maximum moments will give us information only on periodicity of a phenomenon and its variations, it will not suffice. To solve the problem of Blazhko-effect and related problems, synchronous photoelectric multicoloured observations are ncessary as well as determinations of radial velocities. Only large instruments can solve these problems.

Multiple Periodic Variable Stars
IAU Colloquium No.29, Budapest, 1975

OBSERVATIONAL ASPECTS OF MULTIPERIODICITY IN THE β CANIS MAJORIS STARS

Janet Rountree Lesh, Department of Physics and Astronomy,
University of Denver

Morris L. Aizenman, Joint Institute for Laboratory Astrophysics, National
Bureau of Standards and University of Colorado

I. Introduction

Multiperiodicity has been recognized as an important aspect of β Canis Majoris-type variability ever since the intensive observations of the "classical" members of this group by Struve and his coworkers in the early 1950's. About half the stars observed by Struve exhibited a radial-velocity variation whose period was several hours, but whose amplitude was modulated with a period of the order of days. This quickly became known as the "beat phenomenon," and was interpreted as an interference between two nearly equal short periods.

The number of stars considered to be certainly or very probably members of the β Canis Majoris class has roughly doubled since the early 1950's (although the number of stars involved remains relatively small), but the percentage of variables having two or more short periods has remained about the same. Among newly discovered as well as among "classical" variables, about half the stars seem to have only <u>one</u> short period, with no periodic modulation. In the present discussion, we shall call these the Class I variables. In a few additional cases, all involving members of binary systems, the one short period seems to be superimposed on a longer-period variation that can be attributed to orbital motion. We shall call these systems Class II variables. Finally, the truly multiperiodic variables -- those having two or more short periods, whether or not a binary period is present -- will be called Class III variables.

II. The Observations

The observational data for the Class I stars are given in Table 1, where the stars are listed roughly in the order of decreasing quantity of available information. With the exception of BW Vul, to which we shall return later, these stars all appear to be relatively simple objects (which is not to say that they are simple to explain!). None of them is known to be a binary star. All have a single, well-determined period which, at least in the well-observed cases of δ Cet and ξ^1 CMa, does not show secular variations. The light and velocity curves are nearly sinusoidal, and there are no large changes in the line-width (like those observed by Struve in the multiperiodic variables), nor phase differences in the velocity

Table 1

Single-Period Stars

Star	P(d)	2K (km/s)	Δv (mag)	dP/dt (s/cent)
γ Peg	0.1517495	7.0	0.017	
δ Cet	0.1611380	12.6	0.025	0.0
ξ^1 CMa	0.2095755	36.0	0.034	0.0
BW Vul	0.2010249 (1925)	160 (var)	0.2 (var)	+4
HR 6684	0.1398903	22.6	0.028	
τ^1 Lup	0.177365	10.6	0.027	
V986 Oph	0.2907 ?	33 ?	0.014 ?	
o Vel	0.131977	7	0.03	
δ Lup	0.16547 ?		0.0035	

curves as derived from lines of different elements (van Hoof effect). However, recent high-resolution work such as that of Le Contel (1969) on γ Peg suggests that even these stars may undergo subtle line-profile variations.

BW Vul is an exceptional member of this class, since its period is increasing at a rate of about 4 seconds per century, its velocity amplitude is also secularly increasing, and it shows large spectrum variations. However, since the existence of a single short period seems to be firmly established, we shall include BW Vul in Class I as the exception that proves the rule.

Table 2 gives the observational data for the Class II variables. With the exception of α Lup, which is included in this class by analogy, all are known to be members of binary systems. It seems likely that they would be ordinary Class I variables, were it not for this circumstance. (However, most of them do show small profile variations, and the van Hoof effect has been observed in α Vir and α Lup.) In α Vir, the well-determined orbital period produces light variation as well as velocity varia-

Table 2

Single-Period Stars with Modulation

Star	Short P(d)	2K (km/s)	Δv (mag)	dP/dt (s/cent)	Long P(d)	2K (km/s)	Δv (mag)
β Cep	0.1904881 (1968)	45 (var)	0.0365	+1.2	10.893? 6?	6.6	
α Vir	0.1737853 (1969)	17	0.029 (var)	-1	4.01454	228	0.06
β Cen	0.157	14.4	0		352	32	
λ Sco	0.2137015	17	0.0231		10.1605		0.0079
α Lup	0.259882	20 (var)	0.02 (var)		10 y ?		

tion, because of the ellipticity of the components. There is also long-period light variation in λ Sco, although its period is nearly twice the orbital period. No light variation, of either short or long period, has been observed for β Cen.

The case of β Cep is complicated. Struve considered it to be a single-period variable, despite changes in the velocity amplitude. Long-period variations of 11 days and 6 days have been reported by Fitch (1969) and Fischel and Sparks (1972), respectively. But although β Cep has two known companions, one of them discovered by speckle interferometry, neither of these is close enough to the primary to have a short enough orbital period to produce the observed variations.

The Class III stars are presented in Table 3, where P_I refers to the short-period oscillation with the largest radial-velocity amplitude, P_{II} to the one with the second-largest amplitude, and P_b to the beat period between them. The first five stars in this list were well observed by Struve, who usually determined the principal period and the beat period, and then derived the secondary period from these. Struve's periods have generally been confirmed and refined by later observers, and in some cases new short-period ($P < 1$ day) and/or long-period variations have also been found. These are listed under "Other Short P" and "Other Long P," respectively. The latter may be connected with binary orbital motion. The last six stars in Table 3 have been discovered more recently, and there is usually some uncertainty as to the value of their secondary period, whose existence is inferred from changes in the principal oscillation.

Many of the Class III variables are known binaries (σ Sco, 16 Lac, θ Oph, β Cru, and κ Sco), and Fitch (1969) has suggested that ν Eri and 12 Lac are binaries on the basis of their long-period modulation. It has not been proven that _all_ Class III stars are members of binary systems.

Table 3

Multiperiodic Stars

Star	P_I(d)	2K(km/s)	Δv(mag)	P_{II}(d)	2K(km/s)	Δv(mag)	P_b(d)	Other Short P(d)	Other Long P(d)	dP_I/dt (s/cent)
β CMa	0.25002246 (1940)	10.6	0.0044	0.2513003 (1940)	6.6	0.0210	49.17 (1940)	0.23904		+0.5(P_I) / −0.5(P_{II})
σ Sco	0.2468429 (1960)	100	0.040	0.2396710	15	0.021	8.2		33.1	+2.3
16 Lac	0.169165	29.6	0.05	0.170845	9.0	0.04	17.15		12.097	
ν Eri	0.1735089	49.0	0.114	0.1779	16	0.067	6.9808		15.79	+0.2
12 Lac	0.19308883	37.5	0.078	0.197358	17.5	0.029	8.9252	0.18	25.85	+0.1 / −0.2
								0.16	39	
								0.15	8.876	
KP Per	0.201753	20(var)	0.072	0.1982	5?	0.036	10	0.01		
θ Oph	0.140531	12	0.04				4-6			
β Cru	0.160474	4.4	0.04				6	0.2365072	7-8 y	
								0.121383		
15 CMa	0.184557	6.5	0.02				2	0.20 or 0.17		+20
								0.19296		
κ Sco	0.1998303	5.8	0.0087	0.205430		0.0038	7.3316	0.189512	2.951	
ε Cen	0.169608		0.0084	0.17696 or 0.2150		0.0034				

All the well-observed Class III stars show strong line-width (i.e. profile) variations, and many of them also have a van Hoof effect (among the well-observed stars, only ν Eri and KP Per do not). Secular changes in the principal period, which are rare in the Class I and Class II stars, have been reported for a number of Class III variables -- but the exact value and the interpretation of these changes remain debatable in most cases.

The prototype star, β CMa, is undoubtedly the most unusual object in Class III. Its two short periods can be observed independently, without recourse to the beat period or a periodogram analysis, because one of them (which we have called P_I) carries most of the radial-velocity variation while the other (our P_{II}) carries most of the light variation as well as the line-profile variations. Moreover, Shobbrook (1973a) has shown that these two periods have secular variations with opposite signs. Finally, there is no evidence that β CMa is a member of a multiple system.

Notes and references to the data on individual stars, with some alternative values, are given in the following paragraphs. The authors' original nomenclature regarding the ordering of periods is followed in these notes, rather than the notation of Tables 1-3.

γ Peg. P = 0.1517495d according to Sandberg and McNamara (1960). Le Contel (1969) found line-profile variations, although McNamara (1953, 1956) did not. There is no van Hoof effect according to McNamara (1956). McNamara (1955) gives 2K = 7.0 km/s. Jerzykiewicz (1970) gives ΔV = 0.017 mag. Struve (1955) lists dP/dt as inconclusive.

δ Cet. Jerzykiewicz (1971a) gives P = 0.1611380d, ΔV = 0.025 mag. McNamara (1955) observed 2K = 12.6 km/s, with no profile variations. According to van Hoof (1968), dP/dt = 0.

ξ^1 CMa. Shobbrook (1973b) gives P = 0.2095755d, dP/dt = 0, Δv = 0.034 mag. McNamara (1955) observed 2K = 36.0 km/s, with no profile variations. Van Hoof (1962e) suggested that the observed period is actually a combination of two interference waves, implying that four overtones are excited. However, Shobbrook (1973b) disagrees with this interpretation.

BW Vul. Petrie (1954) found that the period was 0.2010249d in 1928, but it increases at an average rate of about 4 s/cent (Petrie 1954, Odgers 1956, Percy 1971). The velocity amplitude also increases at a rate of 0.7 km/s per year (Petrie 1954); its value was 160 km/s in 1954 (Odgers 1956). There are complicated line-profile variations (Struve 1954, Petrie 1954, McNamara et al. 1955, Odgers 1956). Although Petrie (1954) found no van Hoof effect, McNamara et al. (1955) observed a velocity discrepancy between the hydrogen lines and those of other elements. The light range is 0.162-0.210 mag (Walker 1954a).

HR 6684. This variable was discovered by Jerzykiewicz (1972). Pike (1974) gives 2K = 22.6 km/s. Earlier, McNamara and Bills (1973) had found 2K < 5 km/s. However, the light curve is so stable that there is probably no beat. The period is 0.1398903d, according to Morton and Hansen (1974), who observed Δv = 0.028 mag.

τ^1 Lup. In his discovery paper, Pagel (1956) reported P = 0.177365d and 2K = 10.62 km/s, with no profile variations. Although the residuals were rather large, Pagel attributed them to observational errors rather than to a beat. Watson (1971) observed Δv = 0.027 mag.

V986 Oph. Jerzykiewicz (1975) gives P = 0.2907d, ΔV = 0.014; but this period does not fit all his observations, nor does a hypothetical interference between two short periods. Periods found by other observers are 0.2890d (Lynds 1959), 0.28465d (van Hoof 1967), and 0.2859d (Hill 1967). The radial velocity is known to be variable (Plaskett and Pearce 1931), but no detailed study has been published.

o Vel. The variability of this star was discovered by van Hoof (1972), who found $P = 0.131977d$, $2K \sim 7$ km/s, $\Delta V = 0.03$ mag. Morton and Hansen (1974) state that o Vel probably has a single period.

δ Lup. The discoverer is Shobbrook (1972), who finds $\Delta v = 0.0035$ mag. The most probable value of the period is 0.16547d, but 0.14273d is also a possibility.

β Cep. Gray (1970) gives $P = 0.1904881d$, $\Delta V = 0.0365$ mag. Struve et al. (1953) found $dP/dt = +1.2$ s/cent (not confirmed by Gray), and 2K variable between 18 and 46 km/s. Fitch (1969) found a long-period modulation of $P = 10.893d$ which could be binary motion (with $2K = 6.6$ km/s), but this modulation was also not observed by Gray. Van Hoof (1962c) found 10 overtone periods in β Cep. Fischel and Sparks (1972) observed a 6-day variation of the ultraviolet C IV multiplet, while Goldberg and Walker (1974) observed line-profile variations with the short period. Speckle interferometry (Gezari et al. 1972) shows that β Cep has a companion at 0.3 arcsec with $\Delta m_V = 5$ mag, in addition to its visual companion.

α Vir. The short period is 0.1737853d according to Shobbrook et al. (1972), with $2K \sim 17$ km/s and Δv decreasing from 0.029 mag in 1969 to essentially 0 in 1971. The orbital period of 4.01454d (first described by Struve et al. 1958) is reflected in both the light curve and the velocity curve, with $2K = 248$ km/s, $\Delta v = 0.06$ mag (Shobbrook et al. 1969). Smak (1970) found $dP/dt = -5$ s/cent, but Shobbrook et al. (1972) and Lomb (1975a) favor -1 s/cent. Lomb (1975a) finds a van Hoof effect and profile variations. The other short periods originally proposed by Shobbrook et al. (1972) are now considered statistically insignificant (Lomb 1975a). Dukes (1974) found four short periods, with slightly different frequencies for the light and radial-velocity variations.

β Cen. Lomb (1975b) finds P = 0.157d, 2K = 7.2 km/s. He also observes profile variations. There is no light variation (Breger 1967, Shobbrook 1973b). There is a long period of 352d associated with a velocity range of 2K = 32 km/s (Breger 1967, Shobbrook and Robertson 1968); this could be a binary period. Beta Cen has a visual companion at 1.3 arcsec (Δm = 3) and is also a spectroscopic binary with equal-brightness components (Breger 1967, Hanbury Brown et al. 1974, Lomb 1975b).

λ Sco. Shobbrook and Lomb (1972) give P = 0.2137015d, Δv = 0.0231 mag. They also find the first harmonic, 1/2 P, to be present in the light curves. Lomb and Shobbrook (1975) found the same two periods in the radial velocities, with 2K = 17 km/s and 6.0 km/s, respectively. A long period of 10.1605d is found from the variation in mean magnitude (Δv = 0.0079 mag). But this is probably not the orbital period, for which Shobbrook and Lomb (1972) suggest 5.6d. The intensity interferometer (Hanbury Brown et al. 1974) shows that λ Sco is a binary with equal-brightness components.

α Lup. Pagel (1956) discovered that α Lup varies with a period of 0.259882d. He observed a radial-velocity amplitude of 2K = 14 km/s and no light variation (Δv < 0.005 mag). However, more recent observations place 2K as high as 20 km/s (Milone 1962, Rodgers and Bell 1962) and Δv as high as 0.02-0.03 mag (van Hoof 1965b). Van Hoof (1964, 1965b) suggests a 10-year modulation period. There is an indication of profile variations and van Hoof effect (Rodgers and Bell 1962).

β CMa. Struve (1950a) found P_1 = 0.25002246d with K_1 = 5.8 km/s, P_2 = 0.2513003d with K_2 = 2.0 km/s, and P_b = 49.1695d. P_1 was constant while P_2 (the line-broadening period) was decreasing in both period and amplitude, so that P_b increases. These findings were essentially confirmed by Struve et al. (1954) and Milone (1965). Shobbrook (1973a) found

P_1 = 0.25003d, P_2 = 0.2512985d, and P_3 = 0.23904d (as well as 1/2 P_1 and 1/2 P_2) in both radial-velocity and photometric data. In his most recent data, the ranges were 0.0044 mag and 10.6 km/s for P_1, 0.0210 mag and 6.6 km/s for P_2, and 0.003 mag and 2.2 km/s for P_3. Shobbrook found that P_1 increases at a rate of 0.5 s/cent while P_2 decreases at the same rate, so that P_b had increased to 49.20d by 1941. Van Hoof (1962d) found the fundamental radial mode, two overtones, and interference oscillations in Struve's data. Line-profile variations were observed by Struve (1950), equivalent-width variations in the Balmer lines by Kupo (1965), and the van Hoof effect by van Hoof and Struve (1953).

σ Sco. Struve *et al*. (1955) found P_2 = 0.246844d, $2K_2$ = 100 km/s; P_1 = 0.255 or 0.239d, $2K_1$ = 15 km/s; P_b = 8.0d. They computed that P_2 was increasing at a rate of 2.3 s/cent. Line-profile variations were observed by Levee (1952) and Huang and Struve (1955); the profiles varied with period P_2. Struve *et al*. (1955) also observed a van Hoof effect. Struve *et al*. (1961) give P_2 = 0.2468429d, P_b = 8.0d, and an orbital period (from the variation of the mean velocity) of 33.19 or 34.23d. Van Hoof (1966) found P_2 = 0.2468406d, with a light range of 0.040 mag; P_1 = 0.2396710d with a light range of 0.021 mag; and P_b = 8.252d. He did not confirm Struve's conclusion regarding the secular increase of the period. He computed the orbital period as 33.008d, exactly $4P_b$. Fitch (1967) recomputed the orbit and found P_{orb} = 33.13d with K = 34.7 km/s.

16 Lac. The best values of the period and amplitude appear to be P_2 = 0.169165d, K_2 = 14.8 km/s, Δv_2 = 0.05 mag; P_1 = 0.170845d, K_1 = 4.5 km/s, Δv_1 = 0.04 mag; P_b = 17.15d (Struve *et al*. 1952a; Walker 1952a, 1954b; McNamara 1957). There is also a periodic variation of the mean radial velocity with a period of 12.097d (Struve *et al*. 1952a), which does not appear in the photometry (Miczaika 1952, Walker 1952a). Small line-

profile variations were observed by Struve et al. (1952a), and a van Hoof effect by van Hoof et al. (1954). Starting with the principal oscillation (Struve's P_2) and the orbital period, Fitch (1969) computed two more short periods and some linear combinations between the short periods and the orbital period.

ν Eri. From the observations of Struve et al. (1952b) and Walker (1952b), we have P_2 = 0.1735089d, K_2 = 24.5 km/s, Δv_2 = 0.114 mag; P_1 = 0.1779d, K_1 = 8 km/s, Δv_1 = 0.067 mag; P_b = 6.9808d. Line-profile variations were observed by Struve et al. (1952b), Struve and Abhyankar (1955), and Laskarides et al. (1971). The latter authors report that there is no van Hoof effect. Van Hoof (1961a,b) found a fundamental oscillation and four overtones, as well as interference waves, in both the photometric and radial-velocity data. Opolski and Ciurla (1962a), Fitch (1969), and Kameswara Rao (1969) all suggested that the data could be represented by one short-period oscillation (Struve's P_2) combined with a long modulation period of 15.79d. The long period could be an orbital period. Struve (1955) gives dP/dt = +0.2 s/cent.

12 Lac. Struve and Zebergs (1955) give P_2 = 0.19308883d, $2K_2$ = 37.5 km/s; P_1 = 0.197358d, $2K_1$ = 17.5 km/s; P_b = 8.9252d. Barning's (1963) analysis of the photometry from the 1956 international campaign confirmed these two periods and added P_3 = 0.182127d and P_4 = 25.85d. Other short periods suggested have been 0.1558 and 0.162d (de Jager 1957), and 0.1541, 0.16 and 0.18d (Sato 1973). Sato gives 0.078 mag for the light range of the primary oscillation, and 0.029 mag for the secondary oscillation. He finds that the primary period is decreasing at a rate of 0.2 s/cent, while Struve (1955) wrote that it was increasing at a rate of 0.1 s/cent. Struve (1951) proposed a long period of 39d; Fitch (1969) suggested an orbital period of 8.876d, based on the work of Opolski and Ciurla (1961,

1962b). Line-profile variations were observed by Struve (1951), Struve and Zebergs (1955), and Struve et al. (1957). The van Hoof effect is mentioned by Struve and Zebergs (1955) and Beres (1966).

KP Per. Jerzykiewicz (1971b) gives $P = 0.201753d$, and confirms the 10-day beat period of Klock (1965). Klock found the secondary period to be 0.1982d with a light range of 0.036 mag, while the light range of the principal oscillation was 0.072 mag. Klock also suggested a 10-15 min variation which was not confirmed by Jerzykiewicz. Slightly different values of the two periods are given by Joshi (1966). Struve and Zebergs (1959) observed variable line profiles but no van Hoof effect.

θ Oph. $P = 0.140531d$ (van Hoof 1962b, Briers 1967), but van Hoof gives $P_b = 6d$ while Briers gives $P_b = 3.9d$. Van Hoof and Blaauw (1958) found $2K = 12$ km/s for the main oscillation, and observed spectral variations. They state that θ Oph is a member of a spectroscopic binary system. The light amplitude is about 0.04 mag (Briers 1967). Van Hoof (1962b) found the fundamental mode, two overtones, and interference waves to be present.

β Cru. Pagel (1956) discovered the variability with a period of $P = 0.160474d$, a velocity range of $2K = 4.4$ km/s, and a light range of 0.04 mag. His period was confirmed by Heintz (1957), who suggested a long period of 7-8 y. But van Hoof (1962a) considered the fundamental period to be 0.23650d, with Pagel's period being the first overtone. Van Hoof also found the second overtone at $P = 0.121383d$, and a beat period of 6d between the fundamental and the interference oscillation of the fundamental with the second overtone. Hanbury Brown et al. (1974) report that β Cru is a binary with $\Delta m = 2.9$ mag.

15 CMa. Shobbrook (1973b) gives $P_1 = 0.184557d$, with the harmonics $1/2\ P_1$ and $1/3\ P_1$ also present. He finds no other short period in his data, despite the variability of the amplitude. On the other hand, he confirms

the idea that there was a 2-day beat period in van Hoof's (1965a) data, giving P_2 = 0.2033 or 0.1690d in 1963-4, and 0.2037 or 0.1681d in 1965. The radial velocity data of Lynds et al. (1956) yield P_1, 1/2 P_1, and P_3 = 0.19296d in Shobbrook's analysis, where $2K(P_1 + 1/2\ P_1) = 9.7$ km/s, and $2K(P_3) = 4$ km/s. The total light range is 0.01-0.02 mag (Watson 1971, van Hoof 1965a, Walker 1956). Shobbrook (1973b) finds that the principal period is not constant, but the form of its variation is doubtful -- the rate of change could be as high as +20 s/cent.

κ Sco. Lomb and Shobbrook (1975) give P_1 = 0.1998303d with a radial-velocity amplitude of 5.8 km/s, P_b = 7.3316d, P_2 = 0.205430d. Another short period, P_4 = 0.189512, may also be present. They state that κ Sco is a spectroscopic binary. Earlier, Shobbrook and Lomb (1972) had found the same two principal periods (with light ranges of 0.0087 mag and 0.0038 mag, respectively), as well as P_3 = 2.951d. They noted that $P_3 = 2P_b/5$, where $2P_b$ could be a "long period" of 14.74d.

ε Cen. Shobbrook (1972) gives P_1 = 0.169608d and P_2 = 0.17696d, with light ranges of 0.0084 and 0.0034 mag, respectively. An alternative value for P_2 is 0.2150d.

III. Discussion

In Table 3 we adopted the notation P_I and P_{II} to denote the strongest and second-strongest oscillations in the stellar radial velocity. But in the past, different observers have tended to use different nomenclatures which reflected both their method of determining the periods and their interpretation of the periods they found. Struve (1950b), for example, designated as P_2 a short radial-velocity period accompanied by changes in the line profiles; a short radial-velocity period without profile variations was called P_1. Furthermore, he denoted by P_3 the beat period between P_1 and P_2, by P_4 the period of amplitude variation of P_2 (seldom if ever observed), and by P_5 the period of variation in the mean velocity of P_2 (this often turned out to be a spectroscopic binary period). Struve believed that either P_1 or P_2 was the orbital period of a very small satellite, while the other short period was either the rotation period or the pulsation period of the primary star. This theory has been largely discarded, because of the obvious difficulty of maintaining a satellite in orbit with the required period.

Van Hoof, on the other hand, in his long series of articles on individual β CMa stars, proposed that these stars are purely radial oscillators in which the fundamental and several overtones, as well as the interference oscillations between them, are excited. He therefore identified the principal period as P_0, indicating that it was the radial fundamental mode; the other, shorter periods he found were called P_1, P_2, etc. to indicate that they were radial overtones. The interference periods were given labels of the type $P_{i,i+2}$; one of these was most often identified with what we have called the secondary oscillation. Although the periods found by van Hoof sometimes gave plausible values for the ratios of overtone to fundamental periods (P_i/P_0) in a simplified stellar model (e.g., a polytrope with $n = 3$ and $\Gamma = 1.52$), the existence of the "overtone" periods is very dubious because van Hoof did not apply any prewhitening to the data before searching for these small oscillations by a periodogram technique.

Wehlau and Leung (1964), in their paper on the application of the periodogram method to observations of short-period variable stars, point out the problems of "diffraction" peaks (due to the total observing "window") and aliases (due to the spacing of individual observations), which must be minimized by use of an apodizing function and by subtracting out the oscillations already found (prewhitening) before a search for smaller oscillations can be made. Gray and Desikachary (1973) showed how to carry out the prewhitening in the frequency domain.

Fitch (1967) suggested a refinement to the periodogram technique, in which the data are first searched in short segments to find approximate periods, thus reducing the extent of the frequency domain that must be searched at very high resolution with the full data set. In addition, Fitch (1967, 1969) studied a number of β CMa stars that either are known spectroscopic binaries (σ Sco, 16 Lac) or exhibit long-period modulation of the mean light and/or radial velocity of their primary oscillation (ν Eri, 12 Lac, β Cep). He denoted the "binary" period by P_0 and the principal short period by P_1, and expressed all the other periods found by the periodogram technique as linear combinations of these (and sometimes of higher overtones). The rationale for this procedure is the theory that the principal oscillation in a binary β CMa star is perturbed by the variation in tide-raising potential as its companion moves in orbit. Those linear combinations of orbital frequency and pulsation frequency are preferentially excited which have natural resonance with another pulsation overtone. Although this is a possible analysis of the multiple periods in some β CMa stars, especially the known binaries, it is doubtful that this theory can explain all cases of multiperiodicity.

Recent observers have frequently used the least-squares method introduced by Barning (1963), a variation on the periodogram technique in which one minimizes the residuals rather than maximizing the amplitude of the period being sought. The application of this method is described by Shobbrook and his coworkers (Shobbrook _et al._ 1972, Shobbrook and Lomb 1972, Lomb 1975c). Since the least-squares and periodogram methods do not _require_ the determination of a beat or other long period, the current tendency is simply to label periods in order of decreasing amplitude: P_1, P_2, etc., a scheme similar to that adopted in Tables 1-3.

Although some authors occasionally attempt to identify a radial or non-radial mode, fundamental or overtone, on the basis of the velocity-to-light amplitude ratio, such identifications must remain doubtful until a theoretical study is made of the expected values of this ratio for various radial and non-radial modes in realistic stellar models. Moreover, as Watson (1971) has pointed out, the correct observational quantity to use in the comparison is not the observed velocity-to-light amplitude ratio, but the ratio of percentage changes in radius and in luminosity. Derivation of this quantity from the observations requires a precise knowledge of the bolometric correction and its variation over the light cycle.

Finally, although this paper has not attempted to deal with the purely theoretical aspects of multiperiodicity in β CMa stars, we may point out some of the pitfalls lying in wait for the theorist. Any successful theory must explain both the 50% of β CMa stars that are multiperiodic .. . and the 50% that are not! It must explain multiperiodicity both in binary stars ... and in single stars like β CMa itself. And this without introducing any assumptions contrary to the observations, such as excessive rotation speeds. It seems likely that the origin of multiperiodicity is intimately connected with the basic instability mechanism for β Canis Majoris stars, and that until the instability is definitely identified, no final conclusion can be reached concerning multiperiodicity.

References

Barning, F. J. M. 1963, B. A. N. 17, 22.

Beres, K. 1966, Acta Astron. 16, 161.

Breger, M. 1967, M. N. R. A. S. 136, 51.

Briers, R. 1967, I. A. U. Comm. 27, Inf. Bull. Var. Stars, No. 200.

Dukes. R. J. 1974, Ap. J. 192, 81.

Fischel, D., Sparks, W. M. 1972, in The Scientific Results from the Orbiting Astronomical Observatory (OAO-2), ed. A. D. Code (Washington: U. S. Government Printing Office), p. 475.

Fitch, W. S. 1967, Ap. J. 148, 481.

Fitch, W. S. 1969, Ap. J. 158, 269.

Gezari, D. Y., Labeyrie, A., Stachnik, R. V. 1972, Ap. J. Letters 173, L1.

Goldberg, B. A., Walker, G. A. H. 1974, Astron. and Ap. 32, 355.

Gray, D. F. 1970, A. J. 75, 958.

Gray, D. F., Desikachary, K. 1973, Ap. J. 181, 523.

Hanbury Brown, R., Davis, J., Allen, L. R. 1974, M. N. R. A. S. 167, 121.

Heintz, W. D. 1957, Observatory 77, 200.

Hill, G. 1967, Ap. J. Suppl. 14, 263.

Hoof, A. van 1961a, Z. f. Ap. 53, 106.

Hoof, A. van 1961b, Z. f. Ap. 53, 124.

Hoof, A. van 1962a, Z. f. Ap. 54, 244.

Hoof, A. van 1962b, Z. f. Ap. 54, 255.

Hoof, A. van 1962c, Z. f. Ap. 56, 15.

Hoof, A. van 1962d, Z. f. Ap. 56, 27.

Hoof, A. van 1962e, Z. f. Ap. 56, 141.

Hoof, A. van 1964, Z. f. Ap. 60, 184.

Hoof, A. van 1965a, Z. f. Ap. 62, 174.

Hoof, A. van 1965b, Kl. Veroff. Remeis Sternw. Bamberg 4, No. 40, p. 149.

Hoof, A. van 1966, Z. f. Ap. 64, 165.

Hoof, A. van 1967, I. A. U. Comm. 27, Inf. Bull. Var. Stars, No. 232.

Hoof, A. van 1968, Z. f. Ap. 68, 156.

Hoof, A. van 1972, Astr. and Ap. 18, 51.

Hoof, A. van, Blaauw, A. 1958, Ap. J. 128, 273.

Hoof, A. van, Ridder, M. de, Struve, O. 1954, Ap. J. 120, 179.

Hoof, A. van, Struve, O. 1953, P. A. S. P. 65, 158.

Huang, S.-S., Struve, O. 1955, Ap. J. 122, 103.

Jager, C. de 1957, B. A. N. 13, 149.

Jerzykiewicz, M. 1970, Acta Astron. 20, 93.

Jerzykiewicz, M. 1971a, Lowell Obs. Bull. 7, 199.

Jerzykiewicz, M. 1971b, Acta Astron. 21, 501.

Jerzykiewicz, M. 1972, P. A. S. P. 84, 718.

Jerzykiewicz, M. 1975, Acta Astron., 25, 81.

Joshi, S. C. 1966, Z. f. Ap. 64, 518.

Kameswara Rao, N. 1969, P. A. S. P. 81, 359.

Klock, B. L. 1965, A. J. 70, 476.

Kupo, I. D. 1965, Astr. Zhurnal 42, 358.

Laskarides, P. G., Odgers, G. J., Climenhaga, J. L. 1971, A. J. 76, 363.

Le Contel, J.-M. 1969, in Non-Periodic Phenomena in Variable Stars, ed. L. Detre (Budapest: Academic Press), p. 297.

Levee, R. D. 1952, Ap. J. 115, 402.

Lomb, N. R. 1975a, Thesis, University of Sydney.

Lomb, N. R. 1975b, M. N. R. A. S. 172, in press.

Lomb, N. R. 1975c, Astrophys. and Space Sci., in press.

Lomb, N. R., Shobbrook, R. R. 1975, preprint.

Lynds, C. R. 1959, Ap. J. 130, 577.

Lynds, C. R., Sahade, J., Struve, O. 1956, Ap. J. 124, 321.

McNamara, D. H. 1953, P. A. S. P. 65, 144.

McNamara, D. H. 1955, Ap. J. 122, 95.

McNamara, D. H. 1956, P. A. S. P. 68, 158.

McNamara, D. H. 1957, Ap. J. 125, 684.

McNamara, D. H., Bills, R. G. 1973, P. A. S. P. 85, 632.

McNamara, D. H., Struve, O., Bertiau, F. C. 1955, Ap. J. 121, 326.

Miczaika, G. R. 1952, Ap. J. 116, 99.

Milone, L. A. 1962, Boln. Asoc. Argent. Astr. 4, 42.

Milone, L. A. 1965, Z. f. Ap. 62, 1.

Morton, A. E., Hansen, H. K. 1974, P. A. S. P. 86, 943.

Odgers, G. J. 1956, Pub. Dom. Astrophys. Obs. 10, 215.

Opolski, A., Ciurla, T. 1961, Acta Astron. 11, 231.

Opolski, A., Ciurla, T. 1962a, I. A. U. Comm. 27, Inf. Bull. Var. Stars. No. 8.

Opolski, A., Ciurla, T. 1962b, Acta Astron. 12, 269.

Pagel, B. E. J. 1956, M. N. R. A. S. 116, 10.

Percy, J. R. 1971, J. Roy. Astron. Soc. Canada 65, 217.

Petrie, R. M. 1954, Pub. Dom. Astrophys. Obs. 10, 39.

Pike, C. D. 1974, P. A. S. P. 86, 681.

Plaskett, J. S., Pearce, J. A. 1931, Pub. Dom. Astrophys. Obs. 5, 1.

Rodgers, A. W., Bell, R. A. 1962, Observatory 82, 26.

Sandberg, H. E., McNamara, D. H. 1960, P. A. S. P. 72, 508.

Sato, N. 1973, Astrophys. and Space Sci. 24, 215.

Shobbrook, R. R. 1972, M. N. R. A. S. 156, 5P.

Shobbrook, R. R. 1973a, M. N. R. A. S. 161, 257.

Shobbrook, R. R. 1973b, M. N. R. A. S. 162, 25.

Shobbrook, R. R., Herbison-Evans, D., Johnston, I. D., Lomb, N. R. 1969, M. N. R. A. S. 145, 131.

Shobbrook, R. R., Lomb, N. R. 1972, M. N. R. A. S. 156, 181.

Shobbrook, R. R., Lomb, N. R., Herbison-Evans, D. 1972, M. N. R. A. S. 156, 165.

Shobbrook, R. R., Robertson, J. W. 1968, Proc. Astro. Soc. Australia 1, 82.

Smak, J. 1970, Acta Astron. 20, 75.

Struve, O. 1950a, Ap. J. 112, 520.

Struve, O. 1950b, P. A. S. P. 64, 20.

Struve, O. 1951, Ap. J. 113, 589.

Struve, O. 1954, P. A. S. P. 66, 329.

Struve, O. 1955, P. A. S. P. 67, 29.

Struve, O., Abhyankar, K. 1955, Ap. J. 122, 409.

Struve, O., McNamara, D. H., Kraft, R. P., Kung, S. M., Williams, A. D. 1952a, Ap. J. 116, 81.

Struve, O., McNamara, D. H., Kung, S. M., Beymer, C. 1953, Ap. J. 118, 39.

Struve, O., McNamara, D. H., Kung, S. M., Hoof, A. van, Deurink, R. 1954, Ap. J. 119, 168.

Struve, O., McNamara, D. H., Kung, S. M., Kraft, R. P., Williams, A. D. 1952b, Ap. J., 116, 398.

Struve, O., McNamara, D. H., Zebergs, V. 1955, Ap. J. 122, 122.

Struve, O., Sahade, J., Ebbighausen, E. 1957, A. J. 62, 189.

Struve, O., Sahade, J., Huang, S.-S., Zebergs, V. 1958, Ap. J. 128, 310.

Struve, O., Sahade, J., Zebergs, V. 1961, Ap. J. 133, 509.

Struve, O., Zebergs, V. 1955, Ap. J. 122, 134.

Struve, O., Zebergs, V. 1959, Ap. J. 129, 668.

Walker, M. F. 1952a, Ap. J. 116, 106.

Walker, M. F. 1952b, Ap. J. 116, 391.

Walker, M. F. 1954a, Ap. J. 119, 631.

Walker, M. F. 1954b, Ap. J. 120, 58.

Walker, M. F. 1956, P. A. S. P. 68, 154.

Watson, R. D. 1971, Ap. J. 170, 345.

Wehlau, W., Leung, K.-C. 1964, Ap. J. 139, 843.

Discussion to the paper of LESH and AIZENMAN

VALTIER: I would like to know if BW Vul presents changes in amplitude.

LESH: Yes, Petrie and Odgers observed a secular increase of 0.7 km/sec per year in the radial velocity amplitude of BW Vul.

COX: Are you able to divide the β CMa stars into groups on the H-R diagram and position them so that one might tell if single period and multiple period stars are in definitely different evolutionary states?

LESH: This is an interesting idea, but unfortunately there is no obvious separation by spectral type (or, equivalently, by log T_{eff} and M_{bol}) between the single period and multiple period β CMa variables.

LE CONTEL: Could you confirm the fact that the best observed stars are the stars you classify in class III (multiple period stars)?

LESH: No, some of the single period stars, including γ Peg, δ Cet, and ξ^1 CMa, have been very well observed.

BAGLIN: Can you comment on the fact that there seem to be variables and non-variables in the same domain of the H-R diagram?

LESH: Morris Aizenman and I have shown that the "instability strip" for β CMa variables is a region of the H-R diagram that is crossed three times by a massive star in the course of its normal post-main-sequence evolution-once in the core-hydrogen-burning phase, once in the secondary contraction phase, and once in the shell-hydrogen-burning phase. Both variable and non-variable stars are present in the "instability-strip." We believe that the simplest explanation for this phenomenon is to assume that the variable and non-variable stars are in different evolutionary stages - most likely the non-variable stars are in the core-hydrogen-burning phase, while the variable stars are in either the secondary contraction or the shell-hydrogen-burning phase.

MOLNAR: In your Table 2, you give a photometric amplitude due to a binary motion for α Vir and λ Sco. Is this due to an occultation or something else such as reflection?

LESH: This is an aspect effect. That is, it is believed that these stars are tidally distorted. If there is an eclipse in the α Vir system, it is so shallow that it has a negligible effect on the total light variation.

Multiple Periodic Variable Stars
IAU Colloquium No.29, Budapest, 1975

ON EXCITATION MECHANISMS OF PULSATION IN β CEPHEI STARS

Shoji Kato

Department of Astronomy, University of Kyoto, Kyoto

The present state of our knowledge concerning the excitation mechanisms of β Cephei pulsation is reviewed. First we shall discuss whether or not oscillations of β Cephei stars can be explained by the conventional pulsation theory. After showing that the explanation by the conventional theory might be impossible, we shall discuss other specific excitation mechanisms suggested so far.

The first question is whether β Cephei stars are unstable to radial pulsation. This problem was examined by Davey (1973). He found that all the models considered are quite stable. His results are reasonable because the usual envelope ionization mechanism is ruled out in β Cephei stars since their surface temperature is high. The excitation of radial pulsation by the so-called ε-mechanism is also not expected in β Cephei stars, since their masses are too small for this mechanism to work.

The next question is whether β Cephei stars are unstable to non-radial oscillations. This problem is not so simple, because the eigen-functions of non-radial oscillations are rather different from those of radial pulsation. In non-radial oscillations, new type of restoring force (or buoyancy force) works on fluid motions, since the motions have the vortex components. Corresponding to this, new type of modes, namely the gravity modes (g-modes), appear in addition to the pressure modes (p-modes). It is these gravity modes that are interesting from the point of view of pulsational instability, because the spatial behavior of their eigen-functions is quite different from that of the pressure modes. Recently, the pulsational instability of non-radial modes with $\ell = 2$ (of surface harmonics Y_ℓ^m) has been investigated numerically by Chiosi (1974), by Aizenman, Cox and Lesh (1975), and also by Osaki (1975). Chiosi and Aizenman et.al. claim independently that they found pulsational instability against some non-radial gravity modes, either during the overall contraction stage (Chiosi 1974) or during the initial stage of the shell hydrogen burning (Aizenman et. al. 1975). On the other hand, Osaki (1975a,b) insists that there is no pulsational instability in any mode and in any model for a 10 solar mass star.

Before discussing the possible cause of this discrepancy, it will be helpful to mention briefly characteristics of gravity modes of non-radial oscillations. In the μ-gradient zone which appears outside the receding convective core, the medium is stratified highly stably to convection. Thus the local characteristic frequency of the gravity oscillations, namely, the Brunt-Väisälä frequency, is high in that region compared with in the outside region. This suggests that gravity modes of non-radial oscillations are partially trapped in the μ-gradient zone. Osaki (1975a) clearly showed this trapping nature of the gravity modes. The propagation diagram and the phase diagram introduced by Osaki (1975a) are very helpful to understand not only this trapping nature but also general characteristics of non-radial oscillations. In addition to the fact that in the μ-gradient zone the gravity modes have large amplitude due to trapping, it is important to note here that in the μ-gradient zone the gravity modes oscillate rapidly in space, because the Brunt-Väisälä frequency is high there. This situation can be seen clearly in Figure 1, which has been reproduced from Osaki's article (1975b).

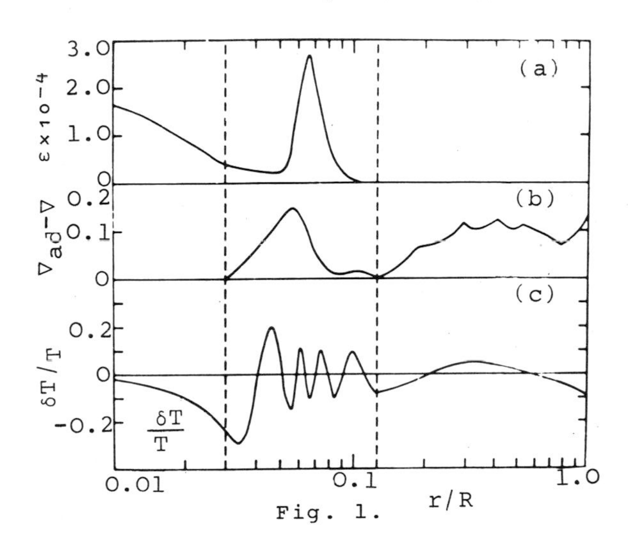

Fig. 1.

Based on the above characteristics of the gravity oscillations, Osaki(1975b) argues that if the μ-gradient zone is treated inaccurately, the radiative damping of oscillations due to rapid spatial variation of temperature in that zone is underestimated. The instability obtained by Chiosi and by Aizenman et.al. might come from this underestimation of the radiative damping. In any way it is difficult to understand why the μ-gradient zone contributes destabilization when the temperature in that zone is stratified sub-adiabatically and there is no shell burning. If we accept Osaki's numerical results and his arguments, we can say that no definite pulsational instability has been found so far within the conventional pulsation mechanism either for radial pulsations or for non-radial oscillations.

Next we shall direct our attention to other specific excitation mechanisms suggested so far. One of them is the "μ-mechanism" proposed by Stothers and Simon (1969). They suppose that the β Cephei stars are the mass-accreting components of close binary systems. By accreting helium-rich material on the envelopes, the reversal of the gradient of mean molecular weight might occur in the outer layer of the stars. This reduces the central condensation of mass, and thus the pulsational amplitude of radial oscillations in the central region increases. The relatively large central pulsation amplitudes permit the nuclear-engined pulsational instability. Stothers and Simon showed that a moderate enrichment of the helium content in the outer envelopes is sufficient to produce instability by this mechanism. This mechanism, however, seems to be unlikely both theoretically and observationally. Arguments concerning (mainly against) this mechanism have been summarized recently in a review article by Cox (1974).

The second mechanism we shall discuss here is that proposed by Osaki (1974). The basic point of this mechanism is the resonant interaction between an overstable convection in a rapidly rotating convective core and a mode of non-radial oscillation, in particular a gravity mode. As is well-known, if a rotation is present a large scale convective motion becomes oscillatory by the action of restoring force of vortex lines. Osaki considered, for example, a star which deviates slightly from a uniform rotation with a faster core at the zero age main sequence stage and evolves from the stage conserving angular momentum in each shell. He found that the frequency of the oscillatory convection of $\ell = m = 2$ actually coincides with that of a gravity mode and thus both modes interact resonantly. Furthermore, estimating the coupling of the two modes, Osaki demonstrated that the amplitude of the non-radial mode to be excited is consistent with the observations.

Concerning Osaki s mechanism, there are some points to be discussed here. The first point is whether the large scale convective motion such as $\ell = m = 2$ is overstable. The convective motions are derived by buoyancy force due to a super-adiabatic temperature gradient, while the restoring force resulting from rotation acts against this. So long as the latter effect of rotation is minor, the convection grows oscillatorily (i.e., overstable). However, if the latter effect of rotation surpasses the former, the adiabatic convective modes become purely oscillatory, not overstable. Osaki shows that the critical super-adiabatic temperature gradient necessary for the convective mode of $\ell = m = 2$ to be overstable is $\nabla - \nabla_{ad} \sim 10^{-3}$, which is rather larger than $\nabla - \nabla_{ad} \sim 5.7 \times 10^{-7}$ expected from the mixing length theory. This gives rise to a question that the large scale convective modes might be pure oscillations and could not be excited in the convective core. One possibility to avoid this difficulty is to take account of the effect of dissipative processes. The purely oscillatory convections forced to be so by the effect of rotation can become overstable if thermal conduction is present. In the case of a uniformly rotating homogeneous medium, the condition of this overstability is that the thermometric conductivity κ is larger than the kinematic viscosity ν (e.g., Ledoux 1958). In the convective core κ and ν come mainly from eddy motions and are essentially of the same order of magnitude (e.g., Unno 1970), although κ is supplemented by radiative diffusivity. In any case, since the present stage of knowledge of turbulent transport is unsatisfactory, it will be allowed to suppose that large scale oscillatory convections can be excited (overstable) in the convective core by the presence of thermal conductivity.

The above considerations suggest another possible excitation mechanism of β Cephei pulsations. That is, the observed pulsation is the large scale oscillatory convection itself excited in a rapidly rotating convective core by the presence of thermal (eddy plus radiative) conductivity. The convective motions certainly can not penetrate till the surface of the star in the case of no rotation. In the present case, however, convections are oscillatory (it can be said that convections are inertial waves) by the effect of rotation and the periods of the oscillations are of the order of those of g_+-modes in the case of no rotation. Thus the oscillatory convections will rather penetrate to the surface of the star. The problem is whether such oscillations can be excited in the star as a whole. The author think this possibility is worth studying, because the mechanism is simple.

The second point which should be discussed on Osaki's mechanism is that his mathematical formulation concerning the resonant coupling between an oscillatory convection and a non-radial oscillation is inadequate. If there is no resonance in frequencies, both modes behave independently of each other in the frame-work of linear theory even if a rotation is present: there is no coupling. If a resonance occurs, the eigen-functions of both modes are mixed and the convective mode may penetrate to surface. In the framework of the linear theory, however, this is still a problem of eigen-value. It will be important to reformulate Osaki's mechanism in this line and to make the condition of excitation of oscillations clear.

The other mechanism which has been suggested so far is a resonant instability of the non-linear coupling between non-radial oscillations and the tidal wave, proposed by Kato (1974). This mechanism is based on the assumption that a β Cephei star is one component of a binary system, and the star has two pulsation modes with very close frequencies. More definitely, it is assumed that the beat frequency f_b of the two pulsation modes is very close to the frequency f_t of the tidal wave induced on the star by the revolution of the secondary star. The frequency of the tidal wave is twice that of the revolution frequency of the secondary. The above condition $f_b \sim f_t$ among frequencies is only a necessary condition. For the resonant interaction to be completed, a spatial coupling between oscillations is also required. Since the tidal wave has two waves in the azimuthal direction, i.e., m = 2, a necessary condtion for the spatial coupling is that the difference of m's of two pulsation modes is two. For example, one mode is the fundamental radial oscillation, and the other is g modes of $\ell = m = 2$. Frequencies of the above two modes are known to coincide actually at an evolutionary stage of β Cephei type stars (e.g., Osaki 1975a). When the above resonant conditions are satisfied, two pulsation modes and the tidal wave interact resonantly. By this resonant interaction energy exchange between pulsation modes and the tidal wave occurs. The direction of energy flow among the waves depends on the forms of eigen-functions. Kato derived the condition where energy is supplied to the pulsation modes from the tidal wave, and thus the pulsation modes grow spontaneously.

The basic assumption included in this mechanism is that the β Cephei stars are components of binary systems as mentioned before. Even if the mass of the secondary star is rather small, pulsation modes seems to be excited if situations are favorable. Thus it is not inconsistent to the fact that all β Cephei stars are not observed as binaries. A weak point of this mechanism is that there is no definite discussion why this mechanism works preferentially at particular evolutionary stages of the stars.

Next we shall comment on the overstability of non-radial modes in the semi-convection zone. Recently, Shibahashi and Osaki (1975) confirmed that non-radial gravity modes are excited in the semi-convection zone, as suggested by Kato (1966). They show that for a mode to be excited, ℓ of the surface harmonics Y_ℓ^m must be large: larger than 15 for a 15 solar mass and larger than 8 for a 30 solar mass. This result can not explain directly the β Cephei phenomena, but non-linear coupling of these overstable modes may excite a mode with small ℓ such as $\ell = 2$. This possibility might be worth studying if all other possible mechanisms failed in explaining the β Cephei phenomena.

Finally we shall note the effects of thermal imbalance. Kato and Unno (1967) and Aizenman and Cox (1975) suggested that the effects of thermal imbalance on non-radial oscillations might play a role in the instability of β Cephei stars. Recently Aizenman, Cox and Lesh examined this problem and found that these effects are much too small to account for the observed instability[1].

In summary, the instability of β Cephei stars seems to be not explained by the conventional pulsation theory, namely, by the theory concerning non-rotating single stars with normal configurations. Some special excitation mechanisms may be required. Some such mechanisms have been proposed, but they are still unsatisfactory and further examinations or refinements are necessary.

To summarize this article, the author had opportunities to talk with Dr. Y. Osaki. The author heartly thanks him for helpful discussion. The author also thanks Prof. W. Unno and Dr. Y. Osaki for comments on the manuscript.

[1] Footnote of the article by Aizenman and Cox (1975).

References

Aizenman,M.L., Cox,J.P. 1975, Astrophys.J., 195, 175.

Aizenman,M.L., Cox,J.P. and Lesh,J.R. 1975, Astrophys.J., 197, 399.

Chiosi, C. 1974, Astron. Astrophys., 37, 281.

Cox,J.P. 1974, Report Prog. Physics, 37, 563.

Davey,W.R. 1973, Astrophys. J., 179, 235.

Kato, S. 1966, Publ. Astron. Soc. Japan, 18, 374.

Kato, S. 1974, Publ. Astron. Soc. Japan, 26, 341.

Kato,S. and Unno,W. 1967, Publ. Astron. Soc. Japan, 19, 1.

Ledoux,P. 1958, Stellar Stability in Handbuch der Physik, ed.S. Flugge (Springer-Verlag, Berlin),p.605.

Osaki,Y. 1974, Astrophys. J., 189, 469.

Osaki,Y. 1975a, Publ. Astron. Soc. Japan, 27, 237.

Osaki,Y. 1975b, Publ. Astron. Soc. Japan, submitted.

Shibahashi,H. and Osaki,Y. 1975, in preparation

Stothers,R. and Simon,N.R. 1969, Astrophys. J., 157, 673.

Unno,W. 1970, Proc. Astron. Soc. Australia, 1, 379.

Discussion to the paper of KATO

HALL: Is there a lower mass limit below which you would not expect the Osaki mechanism to be able to operate?

KATO: Yes, at least the presence of a convective core is necessary. Furthermore, the mass of the convective core must be so large that the excitation of oscillations in the core can surpass their damping in the outer zone.

Multiple Periodic Variable Stars
IAU Colloquium No.29, Budapest, 1975

THE MAGNETIC VARIABLE STARS

Sidney C. Wolff

Institute for Astronomy
University of Hawaii
Honolulu, Hawaii 96822

1. Introduction

The peculiar A stars were first noted as peculiar because they exhibit anomalously strong lines of one or more of the elements Si, Cr, Mn, Sr, and Eu. These diverse objects are all called Ap stars because their assigned spectral types fall predominantly in the range B9-F0. Recent studies have suggested, however, that the Ap stars should be subdivided into two groups. The stars characterized by enhanced lines of Mn, and usually Hg as well, differ from the remaining Ap stars in composition and binary frequency; there are no confirmed spectrum, photometric, or magnetic variations in any Hg-Mn stars; and the upper limit on the magnetic field strengths of the Hg-Mn stars is about 200 gauss. Because of these differences, many observers (e.g. Sargent and Searle 1967; Preston 1971*b*; Wolff and Wolff 1974) have suggested that the Hg-Mn stars are fundamentally different in their properties, and possibly in their origin and evolution, from the SiCrEuSr stars. We shall be concerned today only with this latter group of stars, and in the review that follows only members of the SiCrEuSr class of stars will be referred to as Ap stars.

The problem of the Ap stars, as Bidelman (1967) has stated, is that "stars of unusual spectrum are doing unusual things," and indeed the peculiarities of these objects have been recognized for about three quarters of a century. The star α^2 CVn, which exhibits most of the properties typical of Ap stars, is quite bright (*V* = 2.9) and can be easily observed. The spectrum of α^2 CVn was first classified as peculiar by Maury (1897), who commented on the weakness of the K-line and the strength of the Si II doublet at $\lambda\lambda$4128,31. That this star was of particular interest became apparent when Ludendorff (1906) reported that several lines in the spectrum of α^2 CVn varied in intensity. It is curious, in view of the subsequent analyses of α^2 CVn, that the features reported by Ludendorff to be variable, including lines of Fe, Cr, and Mg, are among the lines that vary least in this star. He noted no variations in the Eu lines λ4129 or λ4205, but it is possible that these lines were outside the range of his instrument.

An extensive analysis of α^2 CVn was carried out by Belopolsky (1913), who showed that the line at λ4129 varied in a period of 5.5 days. This feature, and several other prominent lines in the spectrum of α^2 CVn, were attributed to Eu by Baxandall (1913). Belopolsky (1913) also derived the radial velocities of the Eu lines, and his discussion of the measurements is surprisingly close to the modern interpretation. After demonstrating that the radial velocity of the line at λ4129 varied in quadrature with the changes in intensity, Belopolsky commented (translation by Struve 1942):

> "It is difficult to decide wherein to see the cause of this phenomenon. An obvious hypothesis suggests itself, namely that the central body is surrounded by a gaseous satellite or a gaseous ring having a condensation of matter at one point. This hypothesis is supported by the sign of the variable velocities (negative velocities preceding maximum of intensity of λ4129 and positive velocities following maximum), but the details of the observations still present difficulties which may perhaps be cleared up after more material has been accumulated."

The light curve of α^2 CVn was first measured photoelectrically by Guthnick and Prager (1914). A comparison of their results with a modern light curve (Wolff and Wolff 1971) is shown in Figure 1. The two sets of observations have been phased together according to the period derived by Farnsworth (1933); a slightly better fit could be obtained for a shorter value of the period, but the observations of the Eu variations do not allow such a change.

As Figure 1 shows, Guthnick and Prager not only derived the correct amplitude for the variability but also discovered the asymmetry in the light curve, with the decline from maximum to minimum light occurring more rapidly than the rise from minimum to maximum. The accuracy of the 1914 observations is all the more remarkable since, due to the lack of sensitivity of their equipment, Guthnick and Prager were compelled to use as comparison star δ UMa, which is more than 20° away from α^2 CVn.

Thus by 1914 it was established that α^2 CVn was a spectrum and photometric variable, that the extrema of the light curve coincided in phase with the extrema of the Eu line strength variations, and that the radial

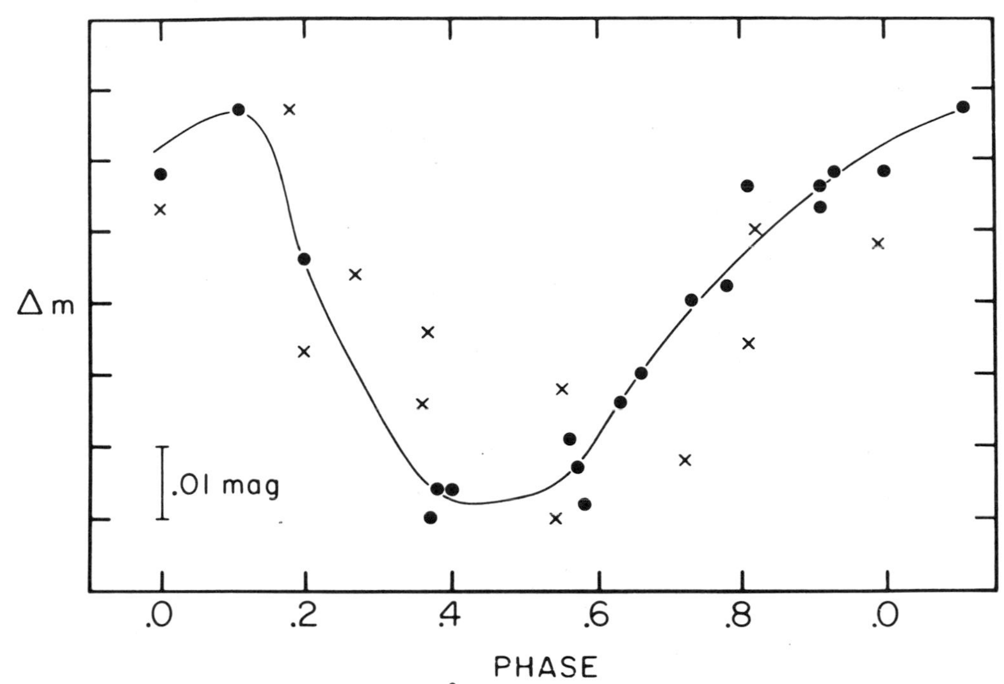

Fig. 1 - Photometric data for α^2 CVn. Crosses represent observations by Guthnick and Prager (1914); filled circles represent observations made with the b filter of the uvby system (Wolff and Wolff 1971).

velocity and spectrum variations were in quadrature. Subsequent studies, particularly by Morgan and Deutsch, showed that these properties are typical of all the Ap variables, although the details of the variations differ from star to star. Modern observations of α^2 CVn itself have been described by Pyper (1969).

2. Rigid Rotator Model

No satisfactory model of the Ap stars was developed until after it was discovered that 78 Vir (Babcock 1947) and most other sharp-lined Ap stars, including α^2 CVn, have variable magnetic fields. The Zeeman observations suggested that a star might possess an axis of symmetry other than the rotation axis, an assumption that is the essential step in formulating the rigid rotator model. This model was first proposed by Babcock (1949) himself:

> "It is true that I have suggested as a revised working hypothesis that intense magnetic activity may be correlated with rapid stellar rotation, but at this stage an equally good case can probably be made for the alternative hypothesis that the spectrum

> variables of type A are stars in which the magnetic axis is more or less highly inclined to the axis of rotation and that the period of magnetic and spectral variations is merely the period of rotation of the star."

Babcock (1960b), of course, was never one of the major proponents of the rigid rotator model. However, none of the alternatives, including the magnetic oscillator and solar cycle models, has been elaborated to the extent that it can successfully predict the variety of observational phenomena associated with the Ap stars. The rigid rotator model, on the other hand, suggests--and has survived--a number of observational tests.

Among the more notable successes of the rigid rotator model are its explanations of the period vs. line-width relation and of the crossover effect and its correct prediction of the average surface field of β CrB. An excellent review of the properties of the Ap stars, together with a discussion of the applicability of the rigid rotator model, has been presented recently by Preston (1971b). In the paper that follows, I will emphasize those results that have been obtained since Preston's review.

3. Light Variations

For a long time it proved impossible to account for the photometric variations of the Ap stars in terms of the rigid rotator model. Indeed, the fact that the periods of the light and magnetic variations are equal has been cited as evidence against the rigid rotator model, since there are no physical arguments to account for the fact that surface brightness appears to depend on the polarity of the magnetic field. However, the discovery that the magnetic fields of several of the Ap stars can be better represented by decentered, rather than centered, dipoles (Wolff and Wolff 1970; Preston 1970; Huchra 1972), indicates that the two magnetic poles are often not of equal strength and that there can be a basic asymmetry between the two magnetic hemispheres.

The key step in understanding the light variations of the Ap stars was the realization (Peterson 1970) that variations in ultraviolet opacity could produce changes in flux in the visible region of the spectrum. In particular, Peterson suggested that in Si variables the continuous opacity in the ultraviolet would be increased at Si maximum, thereby leading to a reduction in flux in the ultraviolet and, due to backwarming effects, to an increase in flux in the visible.

While the backwarming mechanism suggested by Peterson has been basically confirmed as an important cause of the light variations in Ap stars, it is still not clear whether or not Si itself plays a significant role in this process. At the very least, observations demonstrate that there must be other important factors that influence the photometric variations of the Si stars. Several Si stars, including HD 32633 (Preston and Stepien 1968a) and HD 215441 (Babcock 1960a; Preston 1969a) exhibit large amplitude photometric variations even though no Si variations are evident. In 41 Tau (Wolff 1973) maximum light in u (on the uvby system) coincides with Si minimum, a phase relation exactly opposite to that predicted by Peterson's models. In 56 Ari (e.g. Wolff and Morrison 1975) the light curves exhibit two maxima, only one of which coincides with a Si maximum.

As an alternative hypothesis, Wolff and Wolff (1971) suggested that variations in the line opacity due to rare earth elements might be the dominant factor in producing the photometric variations of the cooler Ap stars. Dieke, Crosswhite, and Dunn (1961) have pointed out that there is a great concentration of doubly ionized rare earth lines in the region $\lambda\lambda$ 2000-3000. In late B and early A-type stars, the rare earths should be predominantly doubly ionized, and furthermore a substantial amount of flux is emitted in the region $\lambda\lambda$ 2000-3000. It therefore seemed plausible that variations in the line strengths of the rare earths could directly cause photometric variations. This suggestion received additional support from the fact that in all the rare earth spectrum variables observed up to that time, V maximum coincided with rare earth maximum (Wolff and Wolff 1971; Preston 1971b).

In contrast with the situation a decade ago, predictions about flux distributions below λ 3000 can no longer be made with impunity. Orbiting satellites have made this region accessible to observers, and α^2 CVn was one of the first objects studied in detail by OAO-2. Figure 2 shows the light curves obtained by Molnar (1973) for α^2 CVn. Maximum light in the visible occurs at about phase 0.10 (Wolff and Wolff 1971), so the light variations shortward of λ 2950 are antiphase to the variations longward of this wavelength. Molnar's data also show that the effective temperature of α^2 CVn remains constant throughout the cycle, as would be expected if backwarming effects were responsible for the light variations. Molnar further suggests that rare earth lines are the primary--but not the only--cause of the light variations.

While the photometric variations of most Ap stars can be successfully accounted for by a combination of backwarming from the ultraviolet plus

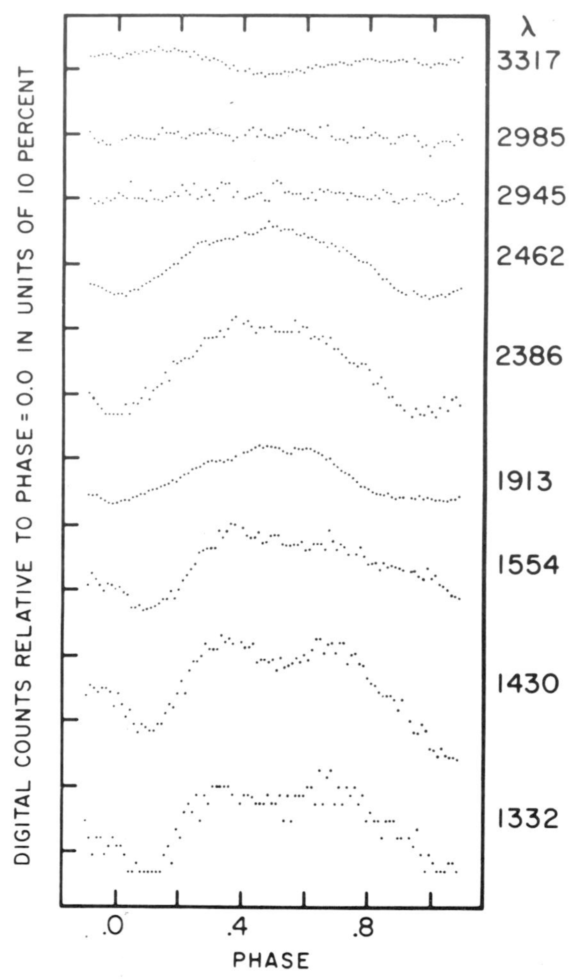

Fig. 2 - Ultraviolet light curves for α^2 CVn (Molnar 1973). Zero phase coincides with Eu maximum.

local line-blocking (e.g. HD 188041; Jones and Wolff 1973), there are several stars whose light curves cannot yet be fully explained. For example, Leckrone (1974) has shown that HD 215441 is similar to α^2 CVn in that the light variations shortward of the null wavelength, which for HD 215441 is at λ2460, are antiphase to the variations longward of this wavelength, and the effective temperature is constant throughout the cycle. The backwarming model thus appears to be applicable to HD 215441, but since no definite spectrum variations have been reported for this star (Babcock 1960a; Preston 1969a), the source of the varying opacity remains unknown. One of the coolest Ap stars, HR 1217, is the only one discovered to date in which V maximum coincides with rare earth minimum (Preston 1972; Wolff and Morrison

1973). The V variation cannot therefore be attributed to backwarming by the rare earths, nor does it appear to be due to changes in line strength within the V passband itself (Bonsack, private communication). In HR 5355 (= HD 125248) the amplitude at λ4100 is substantially larger than the amplitude at λ3600 or λ4600 (Wolff and Wolff 1971; Maitzen and Moffat 1972; Hardorp 1975), and the wavelength variation in the amplitudes cannot be accounted for by backwarming, temperature variations, or local line-blocking (Pilachowski and Bonsack 1975). It may be that there is an unknown source of continuous opacity in the region near λ4100. Some stars, including HD 111133 (Wolff and Wolff 1972; Engin 1974), exhibit large amplitude light variations even though rare earth lines may be weak or absent. Recent work, both theoretical (Leckrone et al. 1974) and observational (Mallama and Molnar 1975) indicate that the lines of the Fe-peak elements may be effective in determining the flux distributions of these objects.

In summary, observational evidence strongly favors the hypothesis that spectrum and photometric variations are causally related. However, several problems must be resolved in order to determine whether or not there are any other causes of photometric variability and before we can explain in detail the light variations of all the Ap stars.

4. Search for Multiple Periods

The model described so far takes into account no periodicities other than the period of rotation, but since this conference is concerned with stars that exhibit multiple periods, I should comment specifically on whether any magnetic Ap stars fall in this category. Before discussing this possibility, I would like to say first that I think recent results indicate that all Ap stars, when observed carefully enough, exhibit cyclic or essentially periodic variability. Periods have now been derived for all but one of the stars that Babcock (1960b) considered to be prototypical irregular variables. Furthermore, the periods of the Ap stars appear to be constant within the present accuracy of measurement (cf. Renson 1972). For example, the observations of α^2 CVn, which span the interval 1913-1970, can all be represented by the period derived by Farnsworth (1933). In some Ap stars (e.g. Bonsack and Wallace 1970), irregular fluctuations may be superposed on cyclic variations. However, the question that concerns us here is whether there are regular variations, possibly due to pulsation, in addition to the periodic variations that are a consequence of the rotation of the star. Several observers have attempted to resolve this question, including particularly Rakos (1963) and most recently Percy (1975). The conclusion appears to be that there is one--and only one--Ap star for which the evidence of multiple periodicity is quite convincing. This star is 21 Com, which has a period of about 30 minutes superposed on a longer period of either 1 or 2 days (Bahner and Mawridis 1957; Percy 1973). In addition, Percy (1975) finds that HR 9080 (=HD 224801) has irregular fluctuations of about $0\overset{m}{.}02$, while 13 other Ap stars show no evidence of short period variability. The incidence of pulsation among the Ap stars that fall in the instability strip appears therefore to be lower than it is for non-Ap stars of the same temperature (Breger 1969). Of the two stars that do show evidence of variability on a short time scale, one (HR 9080) falls clearly outside of the instability strip and the other (21 Com) probably is slightly hotter than the high temperature boundary of the instability strip (Breger 1969).

5. Evolution of Ap Stars

5.1 Introductory Comments

In discussing the Ap stars so far I have essentially ignored variations in their individual properties. With respect to developing a model for their variations, I believe this approach is justified since the rigid rotator model appears to be applicable to all the SiCrEuSr Ap stars. However, I think that the most important unanswered questions concerning the Ap stars deal with their origin and evolution, and here one must be careful not to oversimplify the problem by ignoring the very real differences among various members of this class of stars. A successful explanation for the abundance anomalies, for example, will have to account for the fact that stars of apparently similar temperature and luminosity have quite different compositions. The stars HD 51418 and HR 465, at the time of rare earth maximum, have much stronger lines of Ho and Dy and other heavy rare earths than do most Ap stars. The Ap stars span a large range in temperature, from HR 7129 with T_e = 20000 K (Wolff and Wolff 1976) to HR 1217 (Preston 1972) at T_e = 7000 K. (New observations (Wolff and Hagen 1976) demonstrate that HD 101065 has a strong magnetic field, but the temperature (Wegner and Petford 1974) of this star and its relation to the Ap stars remain matters of controversy.) A successful model will have to explain why magnetic stars occur within, but apparently not outside, of this temperature range.

Another important question concerns the time scale for the development of Ap stars. At different times it has been suggested that these stars are in a variety of evolutionary states from the zero-age main sequence (Hyland 1967) to post-red giant (Fowler et al. 1965) phase of evolution. Recently, new observational evidence has been obtained to support the idea that the Ap stars are on the main sequence for the first time, an idea that has been widely accepted for some time, and that their angular momentum, and possibly other properties as well, change in a systematic way as the stars evolve away from the zero-age main sequence. I would now like to describe the evidence in support of this point of view.

5.2 Distribution of Periods for the Ap Stars

In 1970, Preston and Wolff reported that the Ap star HR 465 was a spectrum, photometric, and magnetic variable with a period of 22-24 years. Subsequent observations (Wolff, unpublished) have confirmed this period. In the interval 1967-1972 the field of HR 465 declined from +200 gauss to -1000 gauss; the average field measured by Babcock (1958) in 1948-49 was about -1100 gauss.

The star HR 465 appears to pose serious problems for the rigid rotator model, since it seems unlikely that any star would have a rotation period that exceeds 20 years. Furthermore the long period of HR 465 is not unique; many Ap stars are known to have periods of several hundred days (Wolff 1975a); γ Equ may have a period of 75 years (Bonsack and Pilachowski 1974). Figure 3 shows the distribution of periods (Wolff 1975a) for Ap stars cooler than about 12000 K. (Hotter Ap stars, which all are classified as Si stars, have been excluded since no Si stars are known to have periods greater than 20 days and only a few have periods greater than 5 days). The distribution of periods is continuous, with no distinct separation between stars with $P < 30$ days, for which the rigid rotator model is applicable, and stars with $P > 30$ days. Furthermore, apart from the time scales of variation, the long period stars are in every way similar to the short period stars. For example, in HR 465, the Eu and Cr lines vary in antiphase, as is typical of CrEu stars; the Eu lines are strongest at V maximum; and the extrema of all the variable quantities coincide in phase. The strength of the magnetic field is similar to that found in the short period stars, and the amplitudes of the spectrum, magnetic, and photometric variability are not unprecedented (Jones, Wolff, and Bonsack 1974). The similarities in the variable properties of the long

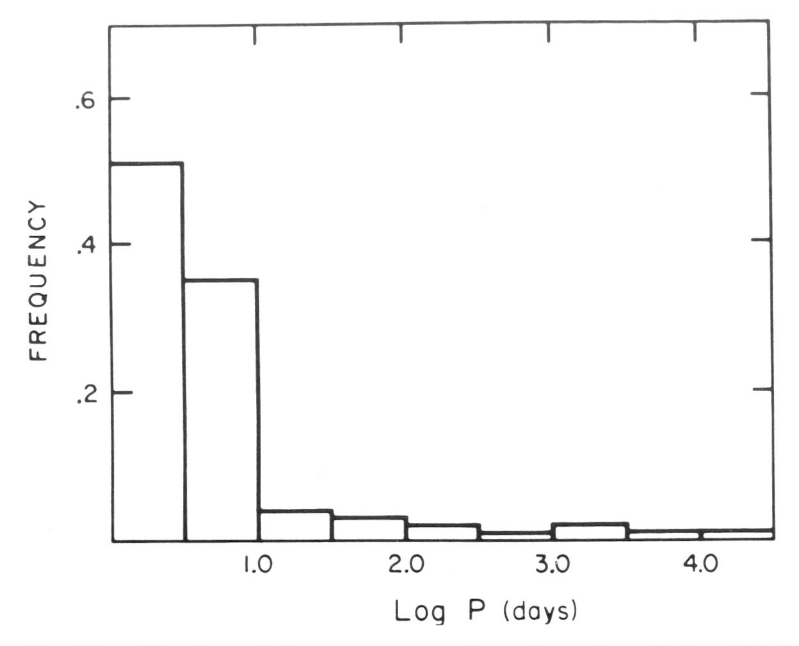

Fig. 3 - Distribution of Ap stars as a function of period. All Si stars have been excluded.

and short period Ap stars, combined with the continuity of the period-frequency distribution, suggest that the same basic mechanism must be responsible for the variations in all Ap stars. Since the rigid rotator model is so successful in accounting for the variations of stars with $P < 30$ days, it is reasonable to ask whether it may also be applicable to the stars of longer period. If it is, of course, then a powerful mechanism for rotational deceleration must be operative in at least some Ap stars.

5.3 Loss of Angular Momentum

The loss of angular momentum of the magnetic Ap stars could occur during either their pre- or post-zero-age main sequence phases. Indeed the fact that Ap stars as a group are slow rotators and few are known with $v \sin i > 100$ km s^{-1} (Abt et al. 1972) suggests that these stars do lose some angular momentum before they reach the main sequence.

Two recent models for the formation of the Ap stars suggest that significant loss of angular momentum can also take place after the magnetic Ap stars reach the zero-age main sequence. One of the models (Havnes and Conti 1971) involves mass accretion from the interstellar medium, the other

(Strittmatter and Norris 1971) postulates mass loss. In each model it is assumed that the magnetic field is strong enough to impose co-rotation out to some limiting radius R_c. In the mass loss model, material crossing this boundary carries angular momentum away from the star. In the mass accretion model, interstellar material crossing the boundary is spun up to the co-rotation velocity, a process that again reduces the angular momentum of the central star. For plausible values of the magnetic field strength and of the density of material at the boundary of the magnetosphere, the e-folding time for loss of angular momentum is on the order of 10^7-10^9 years (Strittmatter and Norris 1971). This time is also comparable to the main sequence lifetimes of the Ap stars.

Decrease of angular momentum through mass accretion is simply the inverse of the process of decrease of angular momentum through mass loss, and accordingly the equations governing the loss of angular momentum are the same for both cases. Thus Havnes and Conti (1971) and Norris and Strittmatter (1971) both predict that the amount of angular momentum lost will depend on the strength of the magnetic field, the density of the stellar wind (mass loss) or of the interstellar medium (mass accretion), and on the length of time that the star has undergone magnetic deceleration. Observationally, there is no way to determine whether, or by how much, the density of material in a stellar wind or in the surrounding interstellar medium varies from star to star as a function of time. Furthermore, observations do not support the idea that present rotational velocities of the Ap stars depend strongly on magnetic field strength (Preston 1971a). For stars with $P > 5$ days, for which accurate photographic measurements of the Zeeman effect can be made, there is no evidence for a correlation between magnetic field strength and rotational velocity (or, equivalently, period).

The remaining observational test of these two models of angular momentum loss is to determine whether the period of a magnetic Ap star depends on its age. If angular momentum is lost during main sequence evolution, then one might expect more evolved stars to have longer periods. In most cases, of course, there is no way to estimate the age of an individual field star. For Ap stars, however, such an estimate is possible. According to Iben's (1965; 1966) evolutionary tracks, as a star in the temperature range of the Ap stars evolves away from the main sequence, it initially increases in radius by nearly a factor of 2 before undergoing the rapid overall contraction that immediately precedes the disappearance of the convective core. Therefore, for stars of a given mass, the radius of the star is a measure of its age. For Ap stars, if the period of variation is equal to the period of rotation, as the oblique rotator model requires, then

$$R = Pv/50.6, \quad (1)$$

where *R* is the stellar radius in units of the solar radius, *P* is the period in days, and *v* is the rotational velocity in km s^{-1}. Since stars are observed at an unknown angle *i*, only *v* sin *i* can be observed, and this equation takes the form

$$R \sin i = (Pv \sin i)/50.6. \quad (2)$$

Tables 1 and 2 list the Ap stars with known *v* sin *i* (Preston, private communication) and with at least moderately reliable values of $P < 30$ days. (For stars with $P > 30$ days, the maximum value of *v* sin *i* is about 5 km s^{-1}, which is below the resolution of most spectrograms obtained to date. A summary of results for Ap stars with $P > 30$ days has been given (Wolff 1975*a*) elsewhere.) Table 1 is restricted to stars for which the same period has been obtained by more than one observer or for which a single observer has obtained the same period for at least two of the three (spectrum, magnetic, and photometric) kinds of variability. Table 2 includes those stars for which periods have been obtained by only a single observer but for which the amplitudes are large enough that the period is probably correct. Tables 1 and 2 do not include all the stars with published periods. For example, I have excluded all stars for which periods have been derived solely from photometric observations of light curves with amplitudes of 0.02 mag. or less. I have also eliminated a number of stars with large amplitudes for which a reanalysis (Hagen and Wolff, unpublished) of the data indicates that alternate periods cannot be ruled out.

Despite this fairly conservative approach, the periods in Table 2 do need confirmation. An example of the problems that can arise is given by HR 6958. For this star, Winzer (1974) derived a period of 0.9451 days from photometric variations with an amplitude of 0.03 mag. In contradiction to these observations, Wolff and Morrison (in preparation) find that the brightness of this star is constant but that the magnetic field varies in a period of 10-12 days.

For the purpose of searching for a correlation between *R* sin *i* and *P*, I have followed Preston (private communciation) in assigning the stars in Tables 1 and 2 to three different temperature classes according to their *UBV* colors, corrected for reddening. While the colors may be slightly affected by blanketing (Wolff 1967), the assignment of these stars to the various temp-

TABLE 1

Ap Stars with Confirmed Periods

HD	Name	Period (days)	v sin i (km s^{-1})	R sin i	Temperature Class	Source
4778	HR 234	2.5475	42	2.11	2	Winzer (1974)
10783		4.1327	24	1.96	2	Preston and Stepien (1968c)
15089	ι Cas	1.73873	46	1.58	3	van Genderen (1970)
18296	21 Per	2.88422	22	1.25	2	Preston (1969c)
19832	56 Ari	0.7278925	200:	2.88:	1	Hardie and Schroeder (1963)
22374	9 Tau	10.61	7	1.47	3	Wolff (1975a)
24712	HR 1217	12.448	≤ 6	≤ 1.48	3	Preston (1972)
25354		3.9001	18	1.39	2	Rakos (1962)
25823	41 Tau	7.227	21	3.00	1	Wolff (1973)
32633		6.431	23	2.92	2	Preston and Stepien (1968a)
34452	HR 1732	2.4660	58	2.83	1	Rakos (1962)
49976		2.976	38	2.23	3	Pilachowski et al. (1974)
51418		5.4379	≤ 20	≤ 2.15	2	Gulliver and Winzer (1973)
62140	49 Cam	4.285	30	2.54	3	Bonsack et al. (1974).
65339	53 Cam	8.0278	< 20	< 3.17	3	Preston and Stepien (1968b)
71866		6.80001	17	2.28	3	Preston and Pyper (1965)
90569	45 Leo	1.4450	13	.37	2	Winzer (1974)
98088	HR 4369	5.90513	25	2.92	3	Abt et al (1968)
108662	17 Com	5.0808	22	2.21	2	Preston et al. (1969)
111133	HR 4854	16.31	10	3.22	2	Wolff and Wolff (1972)
112185	ε UMa	5.0887	34	3.42	3	Guthnick (1934)
112413	α^2 CVn	5.46939	24	2.59	1	Farnsworth (1933)
118022	78 Vir	3.7220	10	.74	3	Preston (1969b)
119213	HR 5153	2.451	35	1.70	3	Wolff and Morrison (1975)
124224	HR 5313	0.52067	130:	1.34	1	Deutsch (1952a)

Table 1 (Continued)

HD	Name	Period (days)	v sin i (km s^{-1})	R sin i	Temperature Class	Source
125248	HR 5355	9.2954	< 15	< 2.76	2	Hockey (1969)
125823	a Cen	8.814	18	3.14	1	Norris (1971)
133029	HR 5597	2.8881	20	1.14	2	Winzer (1974)
137909	β Cr B	18.487	$\leqslant 3$	$\leqslant 1.10$	3	Preston and Sturch (1967)
140160	χ Ser	1.59584	68	2.14	3	Deutsch (1952b)
140728	HR 5857	1.3049	75	1.93	2	Winzer (1974)
152107	52 Her	3.9:	24	1.85	3	Wolff and Preston (1976)
153882	HR 6326	6.00925	26	3.09	3	Preston and Pyper (1965)
173650	HR 7058	9.9748	16	3.15	2	Burke et al. (1969)
175367	HR 7129	3.670	28	2.03	1	Wolff and Wolff (1976)
184905		1.855	70	2.57	2	Burke et al. (1970)
196502	73 Dra	20.2754	8	3.21	3	Preston (1967b)
203006	θ^1 Mic	2.1219	48	2.01	2	Maitzen et al. (1974)
215038		2.036	36	1.45	1	Stepien (1968)
215441		9.488	$\leqslant 6$	$\leqslant 1.13$	1	Stepien (1968)
220825	κ Psc	0.5853	34	.39	2	van Genderen (1971)
223640	108 Aqr	3.73	35	2.58	1	Morrison and Wolff (1971)
224801	HR 9080	3.73983	38	2.81	1	Stepien (1968)

TABLE 2

Ap Stars With Periods In Need of Confirmation

HD	Name	Period (Days)	v sin i (km s^{-1})	R sin i	Temp. Class	Source
7546	HR 369	5.229	33	3.41	1	Winzer (1974)
10221	43 Cas	3.1848	28	1.76	2	Winzer (1974)
14392	63 And	1.3040	78	2.01	1	Winzer (1974)
24155	HR 1194	2.5352	52	2.61	1	Winzer (1974)
27309	56 Tau	1.5691	66	2.05	1	Winzer (1974)
43819	HR 2258	1.0785	14	0.30	1	Winzer (1974)
72968	3 Hya	5.57	16	1.76	2	Wolff and Wolff (1971)
115708		5.07	13	1.30	3	Wolff (1975a)
137949	33 Lib	23.26	10	4.60	3	Wolff (1975a)
170000	Φ Dra	1.7164	89	3.02	1	Winzer (1974)
171586		2.1436	39	1.65	3	Winzer (1974)
177410	HR 7224	1.1663	110:	2.54	1	Winzer (1974)
192913		16.498	14	4.56	2	Winzer (1974)
193722	HR 7786	1.13254	40	0.90	1	Winzer (1974)
221394	HR 8933	2.8419	53	2.98	3	Winzer (1974)
216533		17.20	7	2.38	3	Wolff and Morrison (1973)

TABLE 3

Average v sin i for Ap Stars

Temperature	< v sin i > (km s^{-1})
12300 K $< T_e$	53
9600 K $< T_e <$ 12300 K	35
$T_e <$ 9600 K	23

erature classes should be correct in nearly all cases. Temperature class 1 includes Ap stars with T_e > 12300 K, class 2, stars in the range 9600 K < T_e < 12300 K, and class 3, stars with T_e < 9600K; all temperatures are on the Schild, Peterson, and Oke (1971) temperature scale. Temperature class 1 includes essentially all the Si stars, while temperature classes 2 and 3 include SiCr and CrEuSr stars.

The plots of $R \sin i$ as a function of period are shown in Figures 4 and 5. In these figures the upper envelope of the points should correspond to the actual values of R for the most massive stars included. For temperature class 1, the hottest stars (a Cen and HR 7129) have masses of about 5 $M_\odot$, provided Ap stars have normal masses, and the radius of a 5 $M_\odot$ star varies from 2.4 $R_\odot$ on the zero-age main sequence to a maximum of 4.25 $R_\odot$ near the end of the main sequence phase of its evolution (Iben 1966). Most stars in Figure 4 have masses in the range 3-4 $M_\odot$ and maximum radii of 3.5-4.0 $R_\odot$. The radii of the Ap stars with T_e > 12300 K are therefore approximately in accord with evolutionary calculations. Apart from a deficiency of stars with $R \sin i$ < 2.5 $R_\odot$ and P > 3.5 days, which may be due to angular momentum loss during main sequence evolution, there is no clear correlation of $R \sin i$ with P.

The data for the Ap stars with T_e < 12300 K are shown in Figure 5. Here the upper limit on the masses should be about 3 $M_\odot$ and the corresponding variation in radius should be from about 1.7 $R_\odot$ on the zero-age main sequence to about 3 $R_\odot$ near the end of the main sequence phase of evolution (Iben 1965). In Figure 5, although the scatter is significant, as would be expected for stars with a range in masses observed at various inclinations, $R \sin i$ does appear to increase with increasing period up to P = 10 days. The upper envelope of the data points corresponds rather well to the variation in R predicted by the evolutionary calculations. The possibility that the correlation between P and $R \sin i$ might be due to selection effects has been discussed and discounted elsewhere (Wolff 1975*b*). The shape of the curve in Figure 5, namely an initial rise followed by a rather sharp turnover, can be accounted for if the rate of angular momentum loss is constant--or nearly constant--during the main sequence lifetime of the Ap stars (Wolff 1975*b*).

The hypothesis that the Ap stars lose angular momentum during their main sequence lifetimes can explain a number of the properties of these objects in addition to the correlation of radius with period. For example, the variation of radius with age is nearly linear during the main sequence

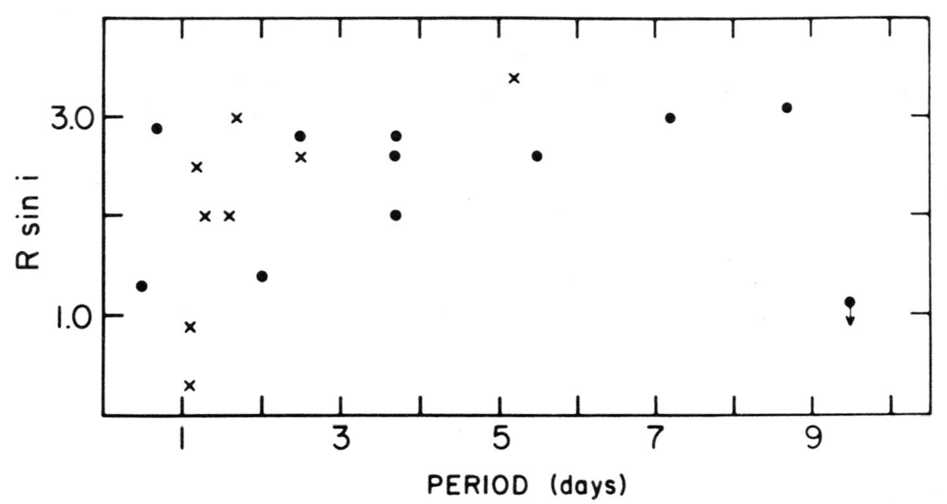

Fig. 4 - Relationship between R sin i, where R is the radius in units of the solar radius, and period for Ap stars with T_e > 12300K. Filled circles represent data from Table 1, crosses represent data from Table 2.

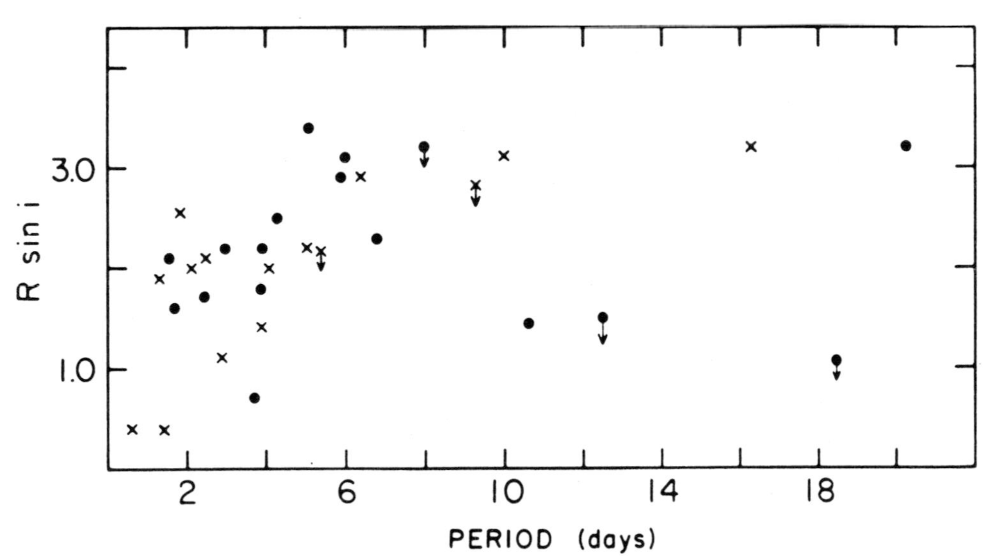

Fig. 5 - Relationship between R sin i and period for Ap stars with T_e < 12300 K. Crosses represent data for stars with T_e in the range 9600 - 12300 K. Filled circles represent data for stars with T_e < 9600 K. All data are from Table 1.

evolution of a 3 $M_\odot$ star. Therefore, the fact that most stars with $P > 6$ days have $R > 2.7\ R_\odot$ suggests that stars lose enough angular momentum to attain such long periods only after completing about 80 per cent of their main sequence evolution. One might therefore expect only about 20 per cent of the Ap stars to have $P > 6$ days, a value roughly comparable to the number actually observed. A more detailed comparison between the expected and observed period-frequency distributions has been carried out by Wolff (1975*b*).

This hypothesis accounts in a natural way for the fact that there are no Si stars with $P > 10$ days and $T_e > 12300$ K. The Si stars are hotter than the stars in Figure 5, their main sequence lifetimes are significantly shorter, and there may therefore be insufficient time to reduce angular momentum to extremely low values. Preston (private communication) has also found that the mean rotational velocities of the Ap stars decrease with decreasing temperature. His results, which are summarized in Table 3, can be explained if angular momentum is lost during the main sequence lifetimes of the Ap stars, and if the time scale for the loss of angular momentum is comparable to the main sequence lifetime of a late B- or early A-type star.

It remains to be seen whether mass loss or accretion can account for the very slow rotation of stars like HR 465. However, since P varies inversely with angular momentum, once the angular momentum is fairly low, then additional small reductions in angular momentum can produce extremely large changes in period (Strittmatter and Norris 1971; Wolff 1975*b*).

5.4 Correlation of Other Properties with Period

If Ap stars do lose angular momentum during their main sequence lifetimes, then the sequence from short to long period stars is an evolutionary sequence, and it is reasonable to ask whether any of the other properties of the Ap stars vary systematically along this sequence.

Recent observations by Landstreet *et al.*(1975) suggest that the strengths of the magnetic fields of Ap stars may be correlated with period. Using photoelectric techniques, Landstreet *et al.* measured magnetic fields in 16 Ap stars with broad lines and found no fields larger than about 1000 gauss; if the distribution of fields for broad-lined stars were like that measured for sharp-lined stars, then Landstreet *et al.* should have detected several stars with fields in excess of 1000 gauss. In interpreting this result Landstreet *et al.* suggest the Ap stars should be divided according to their periods into two groups, those with periods less than 3-5 days having systematically smaller fields than the stars of longer period. They suggest that stars of long period may have become slow rotators precisely because they do have larger magnetic fields, while the rapid rotators were left rotating more rapidly because of their relatively weaker fields. In opposition to this interpretation, it could be pointed out that for stars with $\underline{P} > 5$ days, there is no correlation between period and field strength.

There may be an alternative explanation for the observations of Landstreet et al. Several people (e. g. Mestel 1967; Strittmatter and Norris 1971; Maheswaran 1974) have suggested that magnetic field strengths and rotation may be directly correlated in that, if centrifugal forces due to rotation dominate the magnetic forces, rotational circulation currents may tend to drag the field lines beneath the stellar surface, thus reducing the measured magnetic field. For periods in the range 1-3 days, magnetic and rotational forces are comparable, and it seems possible that rotational circulation currents may reduce the measured magnetic field in these stars, either by pulling field lines beneath the surface or by tangling the field in such a way as to reduce the net longitudinal component of the field, which is all that can be detected by the observational techniques currently in use. If the field is strong enough, however, to result in some magnetic braking, then the magnetic field may come to dominate rotation, with a resultant increase in the measured field strength.

The stars observed by Landstreet et al. differ from the majority of the Ap stars in that they exhibit more rapid rotation and smaller magnetic fields. On the present hypothesis, these stars are also less evolved, and one wonders whether they are in some way less peculiar than Ap stars with longer periods. If, for example, mass accretion were responsible for both the abundance anomalies and the loss of angular momentum in Ap stars, then one might expect abundance and rotation to be correlated. While observations are at present inadequate to determine whether such a correlation exists, there is some information on what kind of correlations cannot exist. First of all, lines of Si, Cr, and Sr are conspicuous even in stars with quite broad lines, so it seems unlikely that the abundances of these elements are strongly correlated with rotation. Detailed abundance analyses of broad-lined Ap stars are, however, not available. Based on measurements of 19 stars, Wolff (1967) found that lines of Eu and other rare earths were weak or absent in stars with $\underline{v} \sin \underline{i} > 40$ km s^{-1}, a result that suggests that rare earth abundances may correlate with rotation. However, subsequent observations have shown that rare earth lines are present in some short period stars (e.g. HD 184905; Babcock 1958; Morrison and Wolff 1971). Furthermore, at least some long period stars (e.g. HD 8441; Babcock 1958; Wolff and Morrison 1973) do not exhibit pronounced rare earth lines. Nevertheless, examination of low dispersion spectral types (Cowley et al. 1969) and of photometric amplitudes of cool Ap stars (Wolff 1975b), which are often correlated with rare earth spectrum variations, suggests that on the average sharp-lined stars may have larger over-abundances of rare earths than do broad-lined stars. Additional observations should be made to determine whether there is indeed a difference of this kind between sharp-lined and broad-lined stars. If there is, then available observations suggest that, as is also true for magnetic field strengths (Landstreet et al. 1975), a period of about 3 days serves to separate the more peculiar stars from the less peculiar ones. There are a number of stars (e.g. 78 Vir and HD 51418) with periods only slightly greater than 3 days that have conspicuously strong lines of the rare earths. Therefore, if there is any correlation between rare earth abundances and rotation, it would have to be in the sense that stars rotating more rapidly than some critical value tend not to show large overabundances, presumably because the atmospheres of stars that rotate more rapidly than this threshold value are not sufficiently stable for the process(es) responsible for forming the rare earth overabundances to be effective.

Analyses of Ap stars in clusters are crucial for determining whether angular momentum, or any other property of these stars, varies systematically during main sequence evolution. Surveys made to date (Young and Martin 1973; Hartoog 1975) indicate that SiCrEuSr Ap stars occur in clusters with only about half the frequency that is thought to obtain for field stars. Such a deficiency of Ap stars in clusters is compatible with the hypothesis that Ap stars develop their peculiarities on a long time scale. However, before accepting this hypothesis as the correct interpretation of the observations, we must know a good deal more about precisely what kinds of Ap stars are found in clusters of various ages.

6. Conclusion

Ten years ago, Preston (1967a) presented a summary of what was then known about Ap stars and suggested several directions for future work. Many of the questions posed by Preston have been essentially answered during the past decade. The number of stars with well-established periods has tripled during that time; it now appears that all SiCrEuSr Ap stars are periodic and that the variations are stable over long periods of time; the evidence in favor of spectroscopic patches (concentrations of specific elements) on the stellar surface seems persuasive (Preston and Sturch 1967; Pyper 1969; Wolff 1969); there appears to be a satisfactory explanation for the light variations of most of the Ap stars; the discovery of resolved Zeeman lines in stars other than HD 215441 (Preston 1969d) has led to much better understanding of the magnetic geometry of the Ap stars (Wolff and Wolff 1970; Preston 1970; Huchra 1972). The extensive effort that has been devoted to understanding the variations of the Ap stars has provided a foundation on which we must now try to build a coherent model of the origin and evolution of the magnetic Ap stars. In the next ten years, I expect that research will increasingly be directed toward answering questions about the time scale for the formation of Ap stars; the source(s) of the abundance anomalies (surely one of the most difficult of the remaining problems); and the relationship of the magnetic Ap stars to normal and non-magnetic peculiar stars.

I am very much indebted to George Preston for making available to me his measurements of rotational velocities for the Ap stars. The preparation of this paper was supported in part by a grant (GP-29741) from the National Science Foundation.

References

Abt, H.A., Chaffee, F.H., and Suffolk, G.: 1972, Astrophys. J. 175, 779.

Abt, H.A., Conti, P.S., Deutsch, A.J., and Wallerstein, G.: 1968, Astrophys. J. 153, 177.

Babcock, H.W.: 1947, Astrophys. J. 105, 105

Babcock, H.W.: 1949, Observatory 69, 191.

Babcock, H.W.: 1958, Astrophys. J. Suppl. 3, 141.

Babcock, H.W.: 1960a, Astrophys. J. 132, 521.

Babcock, H.W.: 1960b, in J.L. Greenstein (ed.), Stellar Atmospheres, Univ. of Chicago Press, p. 282.

Bahner, K., and Mawridis, L.: 1957, Zs. für Astrophys. 41, 254.

Baxandall, F.: 1913, Observatory 36, 440.

Belopolsky, A.A.: 1913, Astron. Nach. 196, 1.

Bidelman, W.P.: 1967, in R.C. Cameron (ed.), The Magnetic and Related Stars, Mono Book Corp, p. 29.

Bonsack, W.K. and Pilachowski, C.A.: 1974, Astrophys. J. 190, 327.

Bonsack, W.K. and Wallace, W.A.: 1970, Publ. Astron. Soc. Pacific 82, 249.

Bonsack, W.K., Pilachowski, C.A., and Wolff, S.C.: 1974, Astrophys. J. 187, 265.

Breger, M.: 1969, Astrophys. J. Suppl. 19, 99.

Burke, E.W., Rice, J.B., and Wehlau, W.H.: 1969, Publ. Astron. Soc. Pacific 81, 883.

Burke, E.W., Rolland, W.W., and Boy, W.R.: 1970, J. Roy. Astron. Soc. Canada 64, 353.

Cowley, A., Cowley, C., Jaschek, M., and Jaschek, C.: 1969, Astron. J. 74, 375.

Deutsch, A.J.: 1947, Astrophys. J. 105, 283.

Deutsch, A.J.: 1952a, Astrophys. J. 116, 536.

Deutsch, A.J.: 1952b, Publ. Astron. Soc. Pacific 64, 315.

Dieke, G.H., Crosswhite, H.M., and Dunn, B.: 1961, J. Opt. Soc. Am. 51, 315.

Engin, S.: 1974, Astron. Astrophys. 32, 93.

Farnsworth, G.: 1933, Astrophys. J. 76, 313.

Fowler, W.A., Burbidge, E.M., Burbidge, G.R., and Hoyle, F.: 1965, Astrophys. J. 142, 423.

Genderen, A.M.: 1970, Astron. Astrophys. Suppl. 1, 123.

Genderen, A.M.: 1971, Astron. Astrophys. 14, 48.

Gulliver, A. and Winzer, J.E.: 1973, Astrophys. J. 183, 701.

Guthnick, P.: 1934, Sitz. Preuss. Akad. Berlin 19, 13.

Guthnick, P. and Prager, R.: 1914, Veröff. Sternw. Berlin-Babelsberg 1, 1.

Hardie, R.H. and Schroeder, N.H.: 1963, Astrophys. J. 138, 350.

Hardorp, J.: 1975, Dudley Obs. Reports No. 9, 467.

Hartoog, M.R.: 1975, Bull. Am. Astron. Soc. 7, 270.

Havnes, O. and Conti, P.S.: 1971, Astron. Astrophys. 14, 1.

Hockey, M.S.: 1969, Monthly Notices Roy. Astron. Soc. 142, 543.

Huchra, J.: 1972, Astrophys. J. 174, 435.

Hyland, A.R.: 1967, in R.C. Cameron (ed.), The Magnetic and Related Stars, Mono Book Corp., p. 311.

Iben, I.: 1965, Astrophys. J. 142, 1447.

Iben, I.: 1966, Astrophys. J. 143, 483.

Jones, T.J. and Wolff, S.C.: 1973, Publ. Astron. Soc. Pacific 85, 760.

Jones, T.J., Wolff, S.C., and Bonsack, W.K.: 1974, Astrophys. J. 190, 579.

Landstreet, J.D., Borra, E.F., Angel, J.R.P., and Illing, R.M.E.: 1975 Astrophys. J., in press.

Leckrone, D.S.: 1974, Astrophys. J. 190, 319.

Leckrone, D.S., Fowler, J.W., and Adelman, S.J.: 1974, Astron. Astrophys. 32, 237.

Ludendorff, H.: 1906, Astron. Nach. 173, 4.

Maheswaran, M.: 1974, Astron. Astrophys. 37, 169.

Maitzen, H.M. and Moffat, A.F.J.: 1972, Astron. Astrophys. 16, 385.

Maitzen, H.M., Breysacher, J., Garnier, R., Sterken, C., Vogt, N.: 1974, Astron. Astrophys. 32, 21.

Mallama, A.D., and Molnar, M.R.: 1975, Bull. Am. Astron. Soc. 7, 270.

Maury, A.C.: 1897, Harvard Annals 28, 96.

Mestel, L.: 1967, in R.C. Cameron (ed.), The Magnetic and Related Stars, Mono Book Corp., p. 101.

Molnar, M.: 1973, Astrophys. J. 179, 527.

Morrison, N.D. and Wolff, S.C.: 1971, Publ. Astron. Soc. Pacific 82, 249.

Norris, J.: 1971, Astrophys. J. Suppl. 23, 235.

Percy, J.R.: 1973, Astron. Astrophys. 22, 381.

Percy, J.R.: 1975, Astron. J., in press.

Peterson, D.M.: 1970, Astrophys. J. 161, 685.

Pilachowski, C.A. and Bonsack, W.K.: 1975, Publ. Astron. Soc. Pacific 87, 221.

Pilachowski, C.A., Bonsack, W.K., and Wolff, S.C.: 1974, Astron. Astrophys. 37, 275.

Preston, G.W.: 1967a, in R.C. Cameron (ed.), The Magnetic and Related Stars, Mono Book Corp., p. 3.

Preston, G.W.: 1967b, Astrophys. J. 150, 871.

Preston, G.W.: 1969a, Astrophys. J. 156, 967.

Preston, G.W.: 1969b, Astrophys. J. 158, 243.

Preston, G.W.: 1969c, Astrophys. J. 158, 251.

Preston, G.W.: 1969d, Astrophys. J. 158, 1081.

Preston, G.W.: 1970, Astrophys. J. 160, 1059.

Preston, G.W.: 1971a, Astrophys. J. 164, 309.

Preston, G.W.: 1971b, Publ. Astron. Soc. Pacific 83, 571.

Preston, G.W.: 1972, Astrophys. J. 175, 465.

Preston, G.W. and Pyper, D.M.: 1965, Astrophys. J. 142, 983.

Preston, G.W. and Stepien, K.: 1968a, Astrophys. J. 151, 577.

Preston, G.W. and Stepien, K.: 1968b, Astrophys. J. 151, 583.

Preston, G.W. and Stepien, K.: 1968c, Astrophys. J. 154, 971.

Preston, G.W. and Sturch, C.: 1967, in R.C. Cameron (ed.) The Magnetic and Related Stars, Mono Book Corp., p. 111.

Preston, G.W. and Wolff, S.C.: 1970, Astrophys. J. 160, 1071.

Preston, G.W., Stepien, K., and Wolff, S.C.: 1969, Astrophys. J. 156, 653.

Pyper, D.M.: 1969, Astrophys. J. Suppl. 18, 347.

Rakos, K.D.: 1962, Lowell Obs. Bull. 5, 227.

Rakos, K.D.: 1963, Lowell Obs. Bull. 6, 91.

Renson, P.: 1972, Astron. Astrophys. 18, 159.

Sargent, W.L.W. and Searle, L.: 1967, in R.C. Cameron (ed.) The Magnetic and Related Stars, Mono Book Corp., p. 209

Schild, R., Peterson, D.M., and Oke, J.B.: 1971, Astrophys. J. 166, 95.

Stepien, K.: 1968, Astrophys. J. 154, 945.

Strittmatter, P.A. and Norris, J.: 1971, Astron. Astrophys. 15, 239.

Struve, O.: 1942, Proc. Am. Phil. Soc. 85, 349.

Tomley, L.J., Wallerstein, G., and Wolff, S.C.: 1970, Astron. Astrophys. 9, 380.

Wegner, G. and Petford, A.D.: 1974, Monthly Notices Roy. Astron. Soc. 168, 557.

Winzer, J.E.: 1974, Ph. D. Thesis, Univ. of Toronto.

Wolff, R.J. and Wolff, S.C.: 1976, Astrophys. J., in press.

Wolff, S.C.: 1967, Astrophys. J. Suppl. 15, 21.
Wolff, S.C.: 1969, Astrophys. J. 157, 253.
Wolff, S.C.: 1973, Astrophys. J. 186, 951.
Wolff, S.C.: 1975a, Astrophys. J., in press.
Wolff, S.C.: 1975b, Astrophys. J., in press.
Wolff, S.C. and Hagen, W.: 1976, in preparation.
Wolff, S.C. and Morrison, N.D.: 1973, Publ. Astron. Soc. Pacific 85, 141.
Wolff, S.C. and Morrison, N.D.: 1975, Publ. Astron. Soc. Pacific 87, 231.
Wolff, S.C. and Preston, G.W.: 1976, in preparation.
Wolff, S.C. and Wolff, R.J.: 1970, Astrophys. J. 160, 1049.
Wolff, S.C. and Wolff, R.J.: 1971, Astron. J. 76, 422.
Wolff, S.C. and Wolff, R.J.: 1972, Astrophys. J. 176, 433.
Wolff, S.C. and Wolff, R.J.: 1974, Astrophys. J. 194, 65.
Young, A. and Martin, A.E.: 1973, Astrophys. J. 181, 805.

Discussion to the paper of WOLFF

WALRAVEN: Would you comment on the variability of Eu in Cepheids?

WOLFF: The fact that Eu is variable in Cepheids as well as in Ap stars was noted several decades ago. In the Ap stars, I think the spectrum variations must be explained in terms of a non-uniform distribution of elements over the surface of a rotating star. The variations in Eu in Ap stars cannot be explained in terms of changes in temperature and pressure. Therefore, I think the variations of Eu in the Cepheids and Ap stars must be due to different causes.

SEGGEWISS: You mentioned that the Ap stars in clusters are only half as frequent as in the field. Is the observational material adequate to justify this statement? I found that there are definitely 30% Ap stars among open cluster blue stragglers (Hg-Mn and Si-Cr-Eu-Sr stars). Another 30% are probable Ap stars.

WOLFF: I based my statement on the work of Young and Martin (1973) and Hartoog (1975). I think that in discussing the frequency of Ap stars in clusters, one should distinguish clearly between the Hg-Mn stars and the Si-Cr-Eu-Sr stars. The time scale for the formation of the Hg-Mn stars is probably $< 10^7$ years (Wolff and Wolff 1974), and so these stars might be expected to occur with the same frequency in clusters as among field stars. If, as I have suggested, the magnetic Ap stars lose angular momentum after reaching the zero-age main sequence, then one might expect that Si-Cr-Eu-Sr stars with very low values of v sin i should be deficient in young clusters.

Multiple Periodic Variable Stars
IAU Colloquium No.29, Budapest, 1975

RED VARIABLES

P. R. Wood

Astronomy Centre, University of Sussex,
Falmer, Brighton BN1 9QH, England.

1. Introduction

This paper will be concerned with those variables lying on the cool side of the cepheid strip in the region of the HR diagram occupied by the red giant and supergiant stars. The blue edge of the red variable region for old disk population stars has been found by Eggen(1973b) to be at $(R-I)_K \sim 0.9$ ($\log T_{eff} \sim 3.53$). According to Eggen(1973a,1973b), all giant stars cooler than this vary, although the amplitude is very small ($\Delta V \sim 0.2$ mag) near the blue edge. In the halo population the blue edge is considerably hotter at $(R-I)_K \sim 0.5$ ($\log T_{eff} \sim 3.62$). On the other hand, some small amplitude red variables (eg. R Dor, AK Hya, IQ Her) have been found at the low temperatures ($R_K - I_K > 1.6$) where large amplitude red (Mira) variables exist. From these results, it appears that the position of the blue edge depends on some unknown stellar parameter(s). Auman(1971) has suggested that the helium abundance may be involved.

The RV Tauri variables, which are heavily reddened by circumstellar dust shells (du Puy1973, Gehrz 1972) and possibly lie in the cepheid instability strip, are not included in the survey.

Before looking at the observational data on red variables, we will look at some theoretical results in order to try to understand the basic pulsation mechanism and to see if any secondary periodicities might be expected.

2. Theory

A distinctive property of highly centrally condensed stars, such as red giants, is the large ratio of the fundamental period P_o to the first overtone period P_1. For a typical red giant star, $P_o/P_1 \approx 3$ to 5, $P_1/P_2 \approx 1.2$ to 1.5 and $P_2/P_3 < 1.2$; these values represent the period ratios that may be expected if two modes are present in a single star.

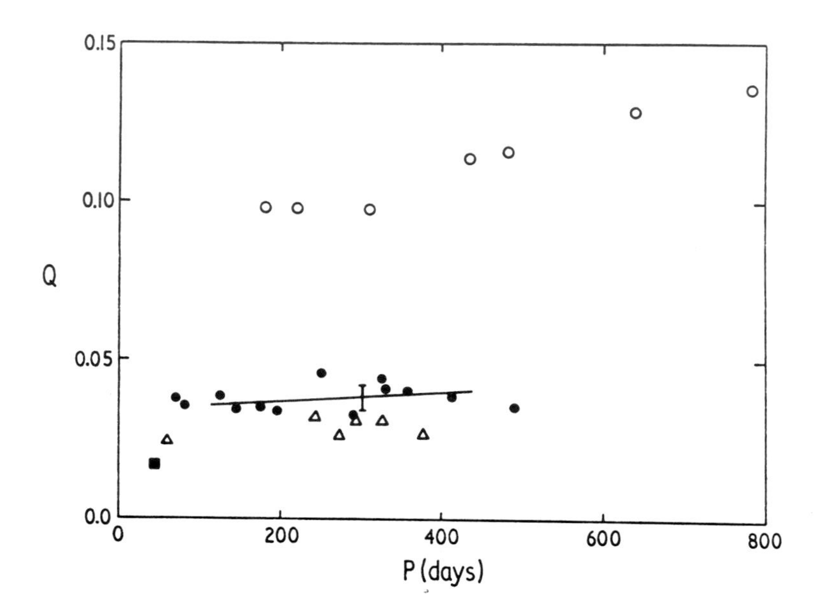

Fig. 1. The pulsation constant Q plotted against period for the fundamental (open circles), first overtone (solid circles) and second overtone (open triangles) modes from theoretical models. The continuous line is the observational result for large amplitude red variables, the error bar indicating a temperature uncertainty of $\pm$ 100°K. The solid square is an average value from observations of small amplitude red variables.

The pulsation constant $Q = P\sqrt{(\frac{M}{M_\odot})/(\frac{R}{R_\odot})^3}$ plotted against period for the first three modes of pulsation from theoretical models by Keeley (1970), Langer (1971) and Wood (1974, unpublished). For such a group of red giants with different central concentrations, the fundamental mode is still well separated from the overtone modes. However, the overtone modes tend to be smeared together.

Full amplitude non-linear models of pulsating red variables have been published by Keeley(1970) and Wood(1974). A general result of these calculations is that as a star evolves up the giant branch and becomes more luminous, it tends to pulsate in a lower order mode. Keeley(1970) found that stable pulsation in both first overtone and fundamental modes was possible. However, in the calculations of Wood(1974,unpublished) the fundamental mode is always found to be very unstable and to have envelope relaxation oscillations superimposed upon it.[1] Model 3 of Wood(1974) is a good example of this type of behaviour. Such pulsation cannot be associated with Mira variables but could possibly be the cause of planetary nebula ejection and the behaviour of some symbiotic stars (Wood 1974).

At lower luminosities on the giant branch, the overtone modes rather than the fundamental mode dominate. Figs. 2 and 3 show light and radial velocity curves for a first overtone pulsator with luminosity $L=4000L_{\odot}$ and a second overtone pulsator with $L=2500L_{\odot}$. Both models have mass $M=1.2M_{\odot}$, composition $(X,Z)=(0.68,0.02)$ and core masses of asymptotic-giant-branch stars. The second overtone pulsator has a smaller light and velocity amplitude than the first overtone pulsator and it also shows a significant cycle-to-cycle variation in the shape of the light curve.

Theoretical calculations show that the luminosity at which the transition between two particular modes occurs in a star increases with the mass of the star. For stars with composition $(X,Z)=(0.68,0.02)$, asymptotic-giant-branch core masses and a mixing-length of one pressure scale height, the transition between first overtone and fundamental pulsation has been found to occur at the following luminosities (Wood, unpublished): $\log\frac{L}{L_{\odot}} = 3.4$ to 3.6 for $M=0.9M_{\odot}$, $\log\frac{L}{L_{\odot}} = 3.6$ to

[1] The difference in the behaviour of the models of Keeley and Wood is probably due to the different treatment of convection and possibly the larger number of mesh points used by Wood.

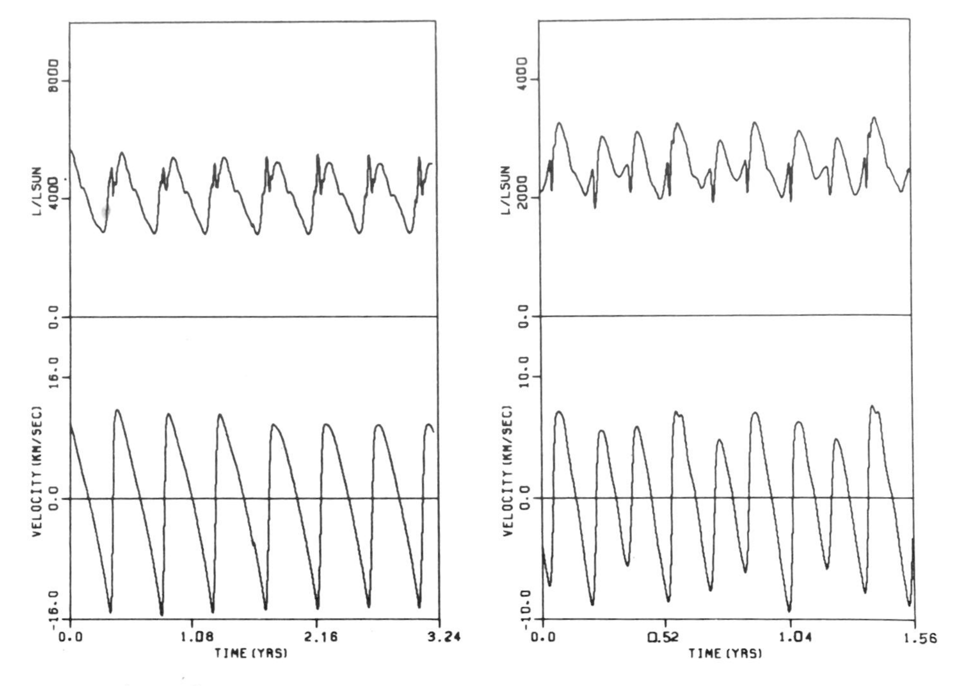

Fig. 2. (left) Light and velocity curves of a theoretical first overtone pulsator.

Fig. 3. (right) Light and velocity curves of a theoretical second overtone pulsator.

3.8 for $M=1.2M_{\odot}$ and $\log\frac{L}{L_{\odot}} > 3.8$ for $M=1.5M_{\odot}$. Although the general trend should remain unaltered, the exact value of the luminosity at which the transition between two given modes occurs will depend on the effective temperature of the model (via mixing-length, composition,etc), as shown by Keeley (1970). These results indicate that, for a given composition, increasing the mass of a star allows it to reach higher luminosities before it switches to fundamental mode pulsation with consequent planetary nebula ejection.

Strong mixtures of modes in theoretical non-linear models of pulsating red giant stars were found by Keeley (1970) in one model. In this model, there was a quasi-periodic switching between fundamental and first overtone pulsation, with a cycle length of ~60 overtone periods. Some of the first overtone pulsators I have studied have a small fundamental component of pulsation present, as seen in Fig. 8 of Wood (1974). The fundamental component causes a small modulation of the first overtone period length and light curve. It is also possible that the apparently irregular component of pulsation noted in the second overtone pulsator shown in Fig. 3 is due to the presence of other modes.

One further type of modulation of the basic pulsation expected in a small number of asymptotic-giant-branch stars is a secular change in period produced by luminosity variations resulting from a shell flash of the helium burning shell. During the flash phase the stars do not deviate from the giant branch except in the last flash cycle when the envelope mass is small (Gingold 1974).

A completely different mechanism for producing irregular and semi-regular variability in red giants has been suggested by Schwarzschild(1975). In Schwarzschild's model, the luminosity variations are caused by surface temperature fluctuations in giant supergranules produced by the extensive convection zones in red giant stars.

In summary, theoretical models predict that the following multiperiodic phenomena may be observed in red variable stars: (a) gradual switching from mode to mode over many pulsation cycles, (b) in first overtone pulsators, the light curve and period length may be modulated by the fundamental mode which has a period 3 to 5 times longer, (c) the period may change continuously due to a change in luminosity resulting from a shell flash in the helium burning shell, and (d) a combination of pulsation and temperature fluctuations in supergranules could lead to luminosity variations with two different timescales.

3. Observations: The basic pulsation mode.

Before looking for multiperiodicities in red variables, we will first try to establish which mode(s) is the source of the primary pulsation. Eggen(1975) has derived the period-luminosity relation $< M_{bol} > = 0.5 mag - 2.25 \log P$

for large amplitude red variables of the Hyades and old disk groups, the giant branches of which lie along the line $< M_{bol} > = -0.65mag - 2.5 < R_K - I_K >$ in the HR diagram. Using these results, Wood(1975) has shown that the large amplitude red variables satisfy the (Q,P) relation given by the continuous line in Fig. 1. The error bar indicates a change of $\pm$ 100°K in the temperature derived from the (R-I, T_{eff}) relation of Johnson(1966). The position of the line in the (Q,P) diagram indicates that the Mira variables are first overtone pulsators. It would be difficult to shift the observed relation to agree with the Q values for fundamental pulsators, but the possibility that second overtone pulsation is dominant cannot be excluded. A further piece of evidence for first overtone pulsation is provided by the halo Mira variable V3 in 47 Tuc for which a value of Q=0.05 is derived from the observations of Eggen(1972,1975), assuming a mass of 0.85 $M_\odot$. The apparent absence of fundamental pulsators among the large amplitude red variables agrees with the theoretical results of Wood(1974), which showed that fundamental pulsators have no stable light curve and are probably associated with planetary nebula ejection and some symbiotic stars. However, the luminosities derived from Eggen's (M_{bol},P) relation ($\log\frac{L}{L_o}$ = 3.92 for a typical Mira with P = 300 days) indicate that the luminosity at which the transition from first overtone to fundamental pulsation occurs in the theoretical models given earlier is too low.

Some features which distinguish first overtone from second overtone theoretical pulsators are the smaller light and velocity amplitudes and the greater apparent irregularity of the second overtone pulsators. The small and intermediate amplitude variables, which lie below the large amplitude variables on the giant branch (eg. Eggen 1971), are much less regular than the large amplitude variables and are probably pulsating in higher overtone modes. An average value of Q (0.017) derived for the small amplitude variables given in Table 1 of Eggen(1973b)is shown by the solid square in Fig. 1 and indicates high overtone pulsation.

Another possible method of distinguishing between first and second overtone pulsation is provided by the position of the humps on the rising branch of the light curve. As shown in Figs. 2 and 3, the humps occur at $<\phi> \approx 0.8$ on the bolometric light curve for first overtone pulsators and at $<\phi> \gtrsim 0.7$ for second overtone pulsators. Observations of the light curves of Mira variables by Lockwood and Wing(1971) at the continuum point at 1.04μ show that humps occur in the light curves of most Mira variables in at least some cycles. The humps generally occur in the expected region but the existing observations are too sparse to allow accurate determinations of the phase of the hump relative to the phase of maximum, both at 1.04μ.

Studies of the kinematics of Mira variables by Feast(1963) showed that the variables with $P<145$ days had systematic motions similar to those with $P \gtrsim 300$ days. On this basis, Feast suggested that the shorter period stars were similar to the most common Miras with $P \sim 300$ days, but pulsating in a higher overtone. A period ratio of 2.4 is predicted in this way, which is incompatible with the two modes being successive overtones but, within the limits of error, could possibly be consistent with the 300 day Miras being fundamental pulsators while the Miras with $P<145$ days are first overtone pulsators. An interpretation of Feast's result which does not require the existence of fundamental pulsation is that the two intermediate period groups $149d < P < 200d$ and $200d < P < 250d$ contain all the halo variables and consequently have higher systematic motions. This is consistent with the observation that the Mira variables in globular clusters generally have periods $P \approx 200$ days (Feast 1972), although two Mira variables of longer period have been identified with the metal rich globular clusters NGC 5927 and NGC 6553 by Andrews et al.(1974). The semi-regular variables in the solar neighbourhood have similar (old disk) systematic motions regardless of period (Feast et al. 1972).

In summary, the observations indicate that the basic pulsation mode of large amplitude red variables is the first overtone. The smaller amplitude semi-regular variables, which lie below the Mira variables on the giant branch, appear to be second or higher overtone pulsators.

4. Observed secondary periodicities.

It has long been recognized that in red variable stars the period length and magnitudes at maximum and minimum vary from cycle to cycle, possibly indicating the presence of secondary periodicities. An analysis of the period length variation in a group of long period variables by Eddington and Plakidis(1929) and Plakidis(1932) led them to conclude that in over 75% of the stars studied, the variations could be explained in terms of (a) errors in the determination of the date of maximum and (b) random and independent deviations of the period length about a mean value. Plakidis(1932) found that of the remaining stars, R Hya and R Aql had continuously decreasing periods while R UMa and R Aur showed a decrease in period in sudden jumps. Sterne and Campbell(1937) looked for period changes in 377 well observed long period variables and found decreases in period of R Hya and R Aql and possibly sinusoidal changes in R Cnc, U Boo and S Ser.

Some examples of O-C diagrams of Mira variables are shown in Fig. 4; the data is taken from the tables of Campbell (1926,1955), together with more recent data kindly supplied by J. A. Mattei of the AAVSO. As well as a random scatter about the calculated date, there appear to be sudden changes in the period, which remains relatively constant before and after the change. Student's statistic

$$t = (\bar{P}_1 - \bar{P}_2)/\sqrt{\left(\frac{1}{n_1}+\frac{1}{n_2}\right)\frac{(n_1-1)\sigma_1^2 + (n_2-1)\sigma_2^2}{(n_1+n_2-1)}}\quad\text{, where}$$

(P_1, n_1, σ_1) and (P_2, n_2, σ_2) are the (mean period, number of cycles, standard deviation of the period) before and after a preselected date, was used to test the significance of apparent period changes in a group of 45 well-observed red variables using data from the above sources. It was found that 14 stars

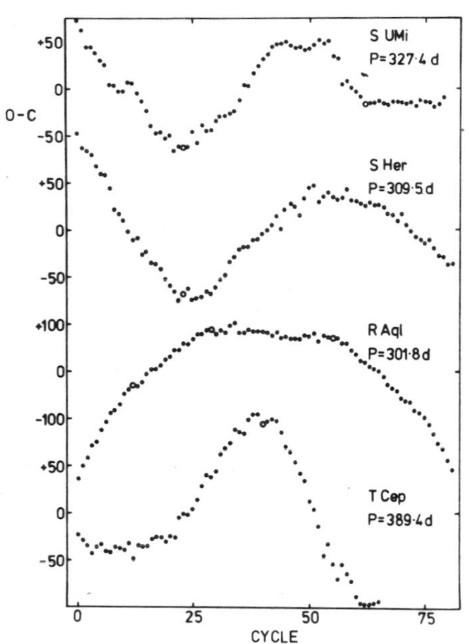

Fig. 4. O-C diagrams for the variables S UMi, S Her, R Aql and T Cep. Open circles indicate period changes at > 99.5% significance as described in the text.

showed changes at the 99.5% significance level with a further 6 having changes at greater than 99% significance. These results indicate that period changes occur in approximately half the red variable stars investigated.[2] Although 14 stars show two or more period changes at greater than 95% significance, the duration of observation is not long enough to show whether these changes represent some quasi-periodic phenomenon or whether the changes occur at random times.

The period changes found at >99.5% significance amount to 2-5% of the period of the star (R Aql excepted), which is too small to be attributed to mode changes. A likely cause of these small period changes is an alteration in the envelope structure near radius $r \gtrsim 0.8R$, which is the region having most influence on the length of the first overtone

---

[2] Preselection of the date at which the test for period change is made is the reason for the large number of significant period changes found here compared with the number found by Sterne and Campbell (1937).

period in red giant stars (Epstein 1950). As the shortest thermal decay timescales for typical red giant envelopes are ≈ 3 years (three or four pulsation cycles), the changes in envelope structure appear to be maintainable over ~ 5 thermal timescales.

An example of a red variable which has probably changed its mode of pulsation is the SRd variable Z Aur, whose period has changed twice from 110.5d to 113.5d and then to 134.8d (Lacy 1973). The latter change, with a period ratio of 1.19, is typical of the period ratio P_2/P_3. This result is consistent with the evidence in the last section which suggested that the small amplitude red variables are pulsating in the second or higher overtones.

The two Mira variables R Hya and R Aql, both of which show a continuous decrease in period, are possible examples of stars undergoing secular luminosity changes resulting from a shell flash in the helium burning shell. Since asymptotic-giant-branch stars do not deviate from the giant branch during a shell flash, the relation between M_{bol} and period (Eggen 1975) given above can be used to calculate the luminosity change in these two stars. Using the periods for R Hya given in the GCVS (Kukarkin et al. 1969) and the periods derived from Fig. 4 for R Aql, the luminosities given in Table 1 result. In R Hya the luminosity is declining on a

Table 1

Star	Date	Period	$\mathrm{Log}\frac{L}{L_\odot}$	Timescale
R Hya	1700	500d	4.117	1200 yrs
	1975	388d	4.018	
R Aql	1915	320d	3.943	550 yrs
	1975	284d	3.896	

timescale $L/\frac{dL}{dt}$ of ~ 1200 years while in R Aql the timescale is ~ 550 years. It is difficult to compare these timescales with theoretical ones because we are not sure of the exact phase of the flash cycle which each of the two stars is in. However, if it is assumed that the stars lie in the phase of luminosity decline immediately following the surface luminosity peak of a flash cycle, then the data of Schwarzschild and Härm(1967) and Sweigert(1971) give timescales $\sim$ 2000 years at $\log\frac{L}{L_\odot} \approx 3.3$ and ~ 50 years at $\log\frac{L}{L_\odot} \approx 4.3$. These theoretical timescales probably represent lower limits as the luminosity decline is relatively rapid in the phase selected. The observed timescales and luminosities are consistent with those derived from the theoretical models. As a further test, the observed fraction of red variables undergoing shell flashes ($\sim 2/400$) can be compared with the theoretically expected fraction. A minimum for the expected fraction f is very roughly the time between the main and secondary peaks of a flash cycle divided by the cycle length. In the models of Schwarzschild and Härm with $\log\frac{L}{L_\odot} < 3.5$, $f \approx \frac{1}{10}$ to $\frac{1}{300}$ while for Sweigert's models with $\log\frac{L}{L_\odot} \approx 4.3$, $f \approx \frac{1}{1000}$. These theoretical values bracket the observed value. It thus appears that R Hya and R Aql are in the active phase of a helium flash cycle; they should be checked for past and present abundance changes such as those found in FG Sge by Langer et al. (1974).

Many semi-regular variables are known to vary with a secondary quasi-period ~ 10 times longer than the primary characteristic period. Some examples of light curves of this type are given for TX Dra and Z Eri by Sacharow(1953), while Payne-Gaposhkin(1954) lists a group of these variables. The N-type carbon star V Hya, for which a long light curve is given by Mayall(1965), may be an extreme example of this phenomenon. The longer of the two periods (typically 500-1500 days) is not inconsistent with this period being the fundamental period of pulsation, which would require the shorter period to be one of the higher overtones ($\sim$5th). Other possible explanations of the longer periodicity are (a) mode switching between overtones in a manner similar to that found by Keeley(1970), (b) some kind of oscillation of the envelope on a thermal timescale which is typically 1000 days, (c) a combination of pulsation and temperature fluctuations in supergranules as suggested by Schwarzschild(1975), and (d) in the carbon stars, periodic grain formation in a circumstellar envelope in a manner similar to that suspected in the R Coronae Borealis stars (Feast and Glass 1973).

Secondary periodicities have been reported in a number of Mira and well-observed SR variables. Van der Bilt (1934) has reported a secondary period of 930 days in the Mira variable SV And, whose primary period (hereafter called P) is 316 days. Fritzová et al. (1954) looked for secondary periodicities in the primary period length and magnitudes at maximum and minimum brightness of a group of 45 long period variables, and found (a) a secondary period ~9 P in V Boo and S Boo (b) a secondary period ~2.7 P in T UMa, and (c) long period (20-50 years) irregular changes in amplitude in R Tri, R Aur, R Cam, T Cas, U Per, S UMa and R Vir. Fischer (1969) has analysed changes in period length (defined in three different ways) in Mira Ceti over 69 cycles using power spectra. Although he found peaks at 2 P and ~10 P in some of his power spectra, his results for different definitions of the period length conflict.

Preliminary results from some calculations of power spectra of the magnitudes at maximum, and of the dates of maximum, are now given. The data used are that described previously. Power spectra (three point running means of the raw power spectra) for three variables in which secondary periodicities have been reported in the literature quoted above are given in Fig. 5. The error bars indicate 99% significance limits for a power spectrum of a white noise process, using a χ^2 test with 6 degrees of freedom. The first ten frequency points in the power spectra of the dates of maximum are joined by dotted lines and should be regarded with caution as the large amount of power generally found in this region is due to the sudden period changes discussed earlier.

In the power spectrum of SV And, there is little evidence for a secondary period at 3 P as suggested by van der Bilt (1934). The light curve of T UMa shows a general excess of power in the region 2-4 times P with possible peaks at 2.2 P and 2.4 P. No peak at ~2.7 P is evident, in disagreement with the

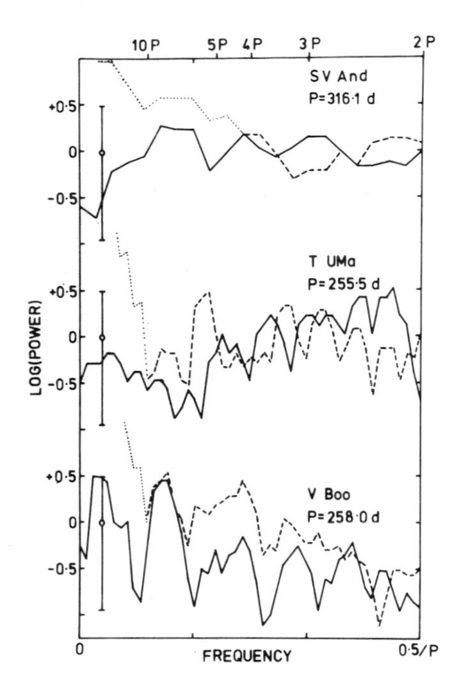

Fig. 5. Power spectra of the magnitudes at maximum (continuous lines) and dates of maximum (dashed lines and dotted lines). Error bars indicate 99% confidence limits.

result of Fritzová et al. (1954). There appears to be a peak in the power spectrum of the dates of maximum of T UMa near 5.5 P. Of the 2 by 46 power spectra calculated, that of V Boo shows the most significant peaks. The two power spectra of V Boo appear well correlated for periods > 3 P and both show a large peak at 8-9 P, in agreement with the secondary period found by Fritzová et al. (1954). No satisfactory explanation of this secondary period is known to me. If it results from beating between two modes then a period ratio of ~1.14 is required, which would require overtones around the fourth to be involved. This appears too high a mode for a star of the amplitude of V Boo.

Some general features which have been noted to occur in the power spectra are (a) strong peaks at 2 P indicating alternating high and low maxima (eg. R And, R Gem), (b) an excess of power in the region 2 to ~ 3 P (eg. T UMa, χ Cyg,

R Cyg), (c) an excess of power at low frequencies in the power spectra of the magnitudes at maximum, indicating long term (20+ years) changes in luminosity (eg. X Cas, T Cas, T Cam), and (d) discrete periods (eg. V Boo). The periods do not appear to occur in both power spectra except in V Boo and probably represent statistical fluctuations. There is no evidence for a predominance of secondary periods in the region 3-5 P which might be expected if the primary period was producing a small modulation of the first overtone. Full results will be published elsewhere when the data has been full analysed.

In summary, the following secondary periodicities have been found in red variables (a) sudden changes in period which may occur on some quasi-periodic timescale in Mira variables, (b) a change in period on a secular timescale in R Hya and R Aql, probably due to a shell flash in the helium burning shell of each of these stars, (c) a quasi-periodic variation in the magnitude of some of the small amplitude red variables on a timescale $\sim$10 P, and (d) a well defined secondary period of 8-9 P in V Boo. There is no general evidence for secondary periods which would occur at 3-5 P if a small amplitude fundamental component of pulsation was superimposed upon the basic first overtone pulsation, as suggested by some theoretical models of Wood (1974, unpublished).

References

Andrews, P.J., Feast, M.W., Lloyd Evans, T., Thackeray, A.D. and Menzies, J.W. 1974, Observatory, 94, 133.

Auman, J.R. 1971, Contr. Kitt Peak National Obs., No. 554, p. 241-246.

Campbell, L. 1926, Harvard Annals, 79, 91.

1955, Studies of Long Period Variables, AAVSO, Cambridge, Mass.

du Puy, D.L. 1973, Astrophys. J., 185, 597.

Eddington, A.S. and Plakidis, S. 1929, Mon. Not. Roy. astr. Soc., 90, 65.

Eggen, O.J. 1971, Astrophys. J., 165, 317.
1972, Astrophys. J., 172, 639.
1973a, Memoirs Roy. astr. Soc., 77, 159.
1973b, Astrophys. J., 184, 793.
1975, Astrophys. J., 195, 661.
Epstein, I. 1950, Astrophys. J., 112, 6.
Feast, M.W. 1963, Mon. Not. Roy. astr. Soc., 125, 367.
1972, in J.D. Fernie (ed.), Variable Stars in Globular Clusters and in Related Systems, D. Reidel, Dordrecht-Holland, p. 131.
Feast, M.W. and Glass, I.S. 1973, Mon. Not. Roy. astr. Soc., 161, 293.
Feast, M.W., Woolley, R. and Yilmaz, N. 1972, Mon. Not. Roy. astr. Soc., 158, 23.
Fischer, P.L. 1969, in L. Detre (ed.), Non-periodic Phenomena in Variable Stars, D. Reidel, Dordrecht-Holland, p. 331.
Fritzová, L., Pěkný, Z. and Švestka, Z. 1954, Bull. astr. Inst. Czech., 5, 49.
Gehrz, R.D. 1972, Astrophys. J., 178, 715.
Gingold, R.A. 1974, Astrophys. J., 193, 177.
Johnson, H.L. 1966, Ann. Rev. Astr. and Astrophys., 4, 193.
Keeley, D.A. 1970, Astrophys. J., 161,657.
Kukarkin, B.V. et al. 1969, General Catalogue of Variable Stars, third edition, Moscow.
Lacy, C.H. 1973, Astr. J., 78, 90.
Langer, G.E. 1971, Mon. Not. Roy. astr. Soc., 155, 199.
Langer, G.E., Kraft, R.P. and Anderson, K.S. 1974, Astrophys. J., 189, 509.
Lockwood, G.W. and Wing, R.F. 1971, Astrophys. J., 169, 63.
Mayall, M. 1965, J. Roy. astr. Soc. Canada, 59, 245.
Payne-Gaposhkin, C. 1954, Harvard Annals, 113, 191.
Plakidis, S. 1932, Mon. Not. Roy. astr. Soc., 92, 460.
Sacharow, G.P. 1953, Variable Stars, 9, 175.
Schwarzschild, M. 1975, Astrophys. J., 195, 137.
Schwarzschild, M. and Harm, R. 1967, Astrophys. J., 150, 961.
Sterne, T.E. and Campbell, L. 1937, Harvard Annals, 105, 459.
Sweigert, A.V. 1971, Astrophys. J., 168, 79.

van der Bilt, J. 1934, Mon. Not. Roy. astr. Soc., 94, 846.
Wood, P.R. 1974, Astrophys. J., 190, 609.
1975, Mon. Not. Roy. astr. Soc., 171, 15p.

Discussion to the paper of WOOD

COX: How do you explain the predicted fundamental mode instability which is not ever observed? How are the observed modes ordered with luminosity, and are they predicted theoretically?

WOOD: If the fundamental mode pulsation leads to planetary nebula ejection, then we don't see the stars as red giants any more since they have become planetary nebula nuclei and nebula shells. If fundamental mode pulsation is associated with symbiotic star behavior, then we do observe fundamental pulsators. At earlier times, when the luminosity was smaller, only overtone pulsation would have occurred.

BATESON: Has any investigation been made of stars at the short end of Mira periods, e.g. T Cen which has large changes in amplitude over many years?

WOOD: No, so far as I know there has been no investigation.

BATESON: In discussing period changes, has attention been paid to the lack of homogeneity in investigations of periods, as investigators use different observational material?

WOOD: Apart from using Campbell's published data, no check has been made on the homogeneity of material.

DZIEMBOWSKI: Do you have any interpretation for those stars that have the period ratios around 9, e.g. V Boo?

WOOD: No.

DZIEMBOWSKI: What is the maximum order of the overtone that may be excited in the red giant region, according to theory?

WOOD: I have calculated only up to the second overtone.

GEYER: In the case of V Hya with a long period variation of 18 years, one should reconsider the old idea of star spot activity. This is also the case where the power spectrum analysis gives a period ratio of 2. Here the stars spots are more concentrated on a part of the stellar hemisphere, and the star rotates with a period of about twice the fundamental pulsation period.

STOBIE: In your models, a large fraction of the luminous flux will be carried by convection. How sensitive do you think your results will be to the theory of the interaction of convection and pulsation which you have used?

WOOD: I think the results are probably qualitatively correct - in particular, the finding that lower order mode pulsation occurs as luminosity increases or temperature decreases. Amplitudes seem to agree with those observed. However, the critical luminosities and temperatures at which the transitions between different modes occur are not reliable.

Multiple Periodic Variable Stars
IAU Colloquium No.29, Budapest, 1975

CEPHEIDS

R.S. Stobie

Royal Observatory Edinburgh, Balckford Hill
Edinburgh EH9 3HJ Scotland.

1. Introduction

The scope of this review will be to summarize our knowledge of Cepheids at the present time concentrating primarily on the observational results obtained but at the same time relating these to the relevant results of pulsation and evolution theory. Of the many previous reviews of pulsating stars we would mention especially Payne-Gaposchkin and Gaposchkin (1938), Ledoux and Walraven (1958), Christy (1966a) and Cox (1974) as providing an excellent background to the theoretical and observational aspects of the subject. Because of these reviews we will concentrate on the more recent results obtained in connection with Cepheids. At the end a special section is devoted to the double-mode Cepheids as they are particularly relevant to this Colloquium.

2. Population I Cepheids

The position of population I or classical Cepheids on the Hertzsprung-Russell (H-R) diagram in relation to other types of intrinsic variable stars is shown in Fig. 1. Population I Cepheids occupy a narrow near vertical strip on the H-R diagram covering the range of absolute magnitude $-2 > M_V > -6$. The stars which occupy this instability strip typically have periods in the range $1 < P < 50$ days, although a few longer period Cepheids do exist.

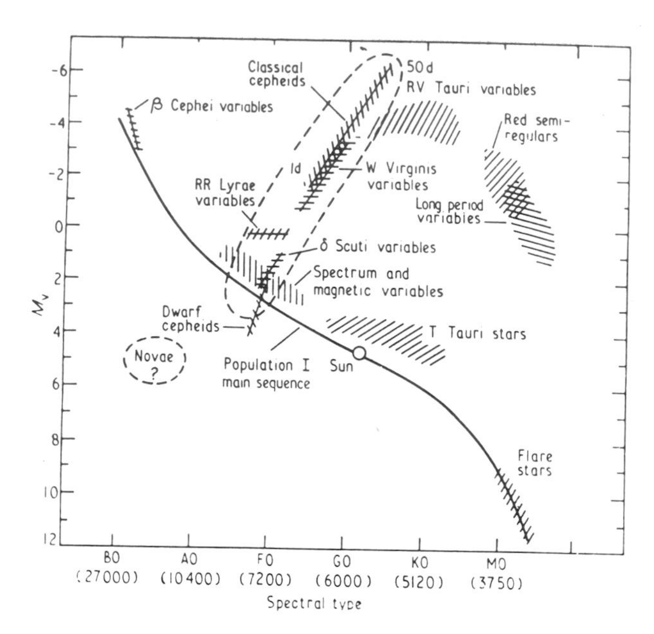

Figure 1. Location of various types of intrinsic variables on the H-R diagram (Cox, 1974).

The evolutionary history of Population I Cepheids is well understood from the extensive series of evolutionary calculations that have been carried out (see, e.g. Iben, 1967). Population I models of mass 4-9 $M_\odot$ as they ascend the red giant branch exhibit loops to the blue at almost constant luminosity on the H-R diagram. These loops can intersect the observed instability strip and this intersection occurs when the models are in the core helium burning stage.

Detailed agreement has been obtained between the theoretical and observed number of Cepheids as a function of absolute magnitude and the period-frequency histogram (Iben 1966, Hofmeister 1967). Differences in the observed short period cut-off between the Cepheids in the Galaxy, the Small Magellanic Cloud (SMC) and the Large Magellanic Cloud (LMC) have led a number of authors to propose that these differences arise as a consequence of the evolutionary loops penetrating the strip to differing degrees in the separate galaxies. The calculations of Robertson (1971) support this suggestion and indicate that both helium content, Y, and heavy element content, Z, affect the loops significantly. The extent of these loops to the blue is sensitive to many factors and extreme care is required in calculating the evolutionary model in order to produce reliable results (Lauterborn, Refsdal and Weigert 1971, Robertson 1971).

Extensive series of linearized, non-adiabatic pulsation calculations have shown the existence of a blue edge to the Population I instability strip which agrees well with the observed blue edge (Iben and Tuggle 1972, King _et al_. 1973). This theoretical blue edge is sensitive both to the helium abundance and to the mass-luminosity (M-L) relationship employed (Fig. 2). A comparison of the blue edges determined by linear and non-linear computations has shown that the blue edges defined by the linear, non-adiabatic calculations are valid as no case has yet been found of hard, self-excited pulsations (King _et al_. 1973). Thus provided an accurate transformation can be made from the observed quantities (M_V, B-V) to the theoretical parameters ($L/L_\odot$, T_e) then this dependence in principle provides a powerful way of determining the helium abundance (assuming some M-L relationship).

The existence of the red edge to the instability strip, however, has not been found theoretically although the suggestion that it is caused by the sudden onset of deep convection (Baker and Kippenhahn 1965) is borne out by recent calculations (Tuggle and Iben 1973).

Theoretical calculations of the periods of pulsating models have all shown that the period is primarily determined by the radius and mass of the model and is almost independent of the luminosity, effective temperature or chemical composition of the model (Christy 1966a, Stobie 1969b, Cox _et al_. 1972a). The period is little affected by non-linearities in the motion and can be accurately determined by calculating models in the linear, adiabatic approximation (Cogan 1970). The existence of this period-radius-mass (P-R-M) relationship together with an M-L relationship from the results of evolutionary models automatically implies that Cepheids will show a P-L-T_e relationship. It is to be

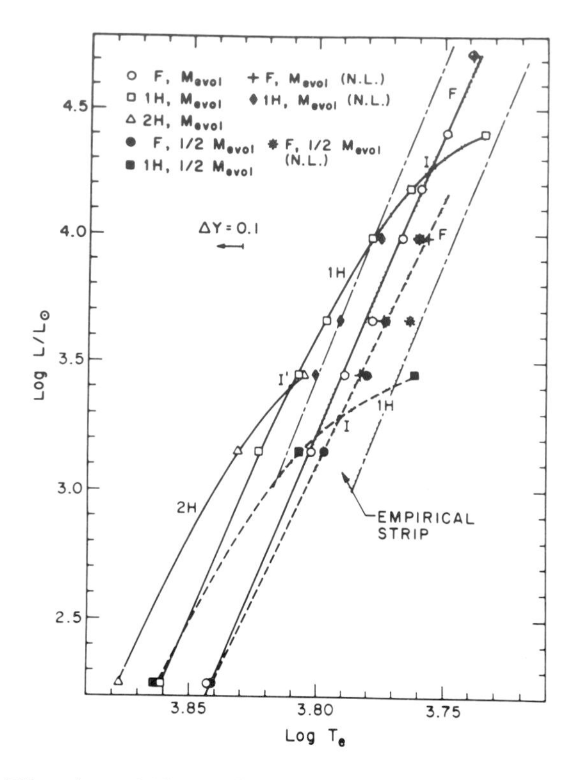

Figure 2. Blue instability edges on an H-R diagram. Solid lines, $M = M_{evol}$; dashed lines, $M = 0.5 M_{evol}$; dashed-dot lines, boundaries of empirical Population I Cepheid strip. All lines in the figure refer to linear, non-adiabatic results and the symbols refer to non-linear, non-adiabatic results. F, 1H, 2H refer to pulsational instability in fundamental mode, first overtone and second overtone respectively. The horizontal arrow shows the estimated shift of the F blue edges that would be brought about by an increase in Y (helium mass fraction) of 0.1 (King et al. 1973).

noted, however, that this P-L-T_e relationship is not independent of chemical composition since the theoretical M-L relationship of Cepheids is dependent upon the chemical composition (Robertson 1971).

The P-L-T_e relationship has found strong confirmation from the observations in the existence of a period-luminosity-colour (P-L-C)

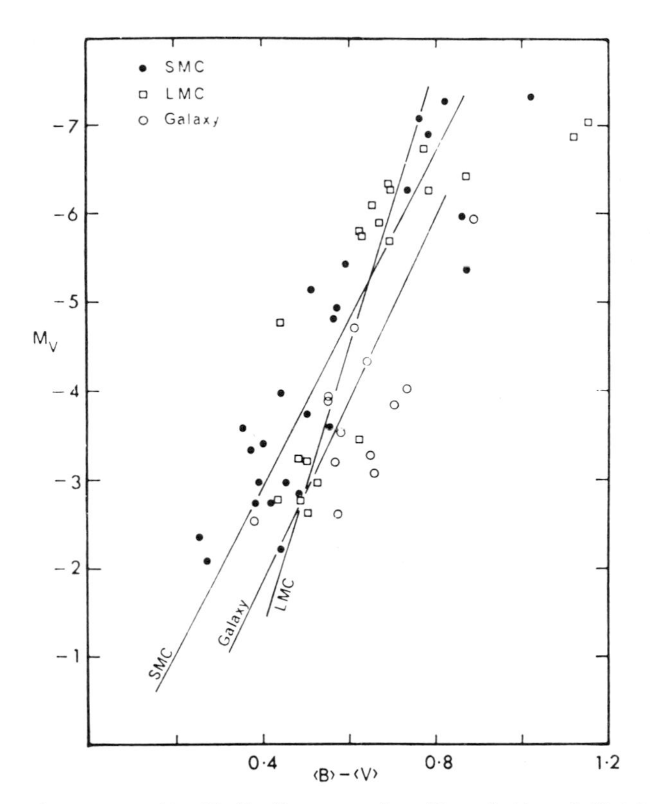

Figure 3. A composite H-R diagram for Population I Cepheids in the SMC, the LMC and our Galaxy (Gascoigne 1969).

relationship. This P-L-C relationship has been determined in our Galaxy from the thirteen calibrating Cepheids (Sandage and Tammann 1969) and also in the SMC and LMC (Gascoigne 1969, van Genderen 1969 and Butler 1975a, b). The relative positions on the H-R diagram of Population I Cepheids belonging to the Galaxy, the SMC and the LMC are shown in Fig. 3. It appears that there are significant differences at the short period end of the instability strip in that the SMC Cepheids are systematically bluer by $0^m.1$ in B-V relative to the LMC and Galactic Cepheids (Gascoigne 1969).

The finite width of the instability strip requires that a third parameter is included in the P-L-C relationship in order to reduce the intrinsic scatter present in the P-L and P-C relationships. Because of the importance of Cepheids as distance indicators it is vital to calibrate the P-L-C relationship as accurately as possible. Sandage and Tammann (1971) have combined the data on Population I Cepheids in different galactic systems to produce a composite and universal P-L-C relationship. On the other hand Gascoigne (1974) has shown that the P-L-C relationship is in fact sensitive to the metal abundance, Z, and he obtains a difference in distance modulus from the SMC of $0^m.41$ depending upon whether the SMC is assumed to have Z = 0.02 or Z = 0.005. Iben and Tuggle (1975) have also presented convincing arguments to show that the P-L-C relationship is not unique from galaxy to galaxy. The basic reason is that if the heavy element abundance varies between different galactic systems this affects the P-L-C relationship by changing both the M-L relationship of Cepheids and the C-T_e transformation (Bell and Parsons 1972). Thus strong doubts have been cast on the applicability of a universal P-L-C relation.

Sandage and Tammann (1971) by combining the data from different galactic systems have studied the amplitudes of Cepheids as a function of position in the instability strip. They find that in the period range $0.40 < \log P < 0.86$ the amplitude decreases monotonically from the blue to the red sides of the strip. Since the amplitude (A) of a variable is found observationally to be a function of its colour and period this means that a P-L-A relationship instead of a P-L-C relationship can be derived for Cepheids in the period range $0.40 < \log P < 0.86$. Subsequent confirmation of this work has been obtained by Kelsall (1972) and Butler (1975c). Madore (1975a, b), however, from a photoelectric study of galactic and extragalactic Cepheids with P > 11 days finds that the amplitudes increase monotonically from the blue to the red sides of the strip in the opposite sense to that described by Sandage and Tammann. Madore also notes that the longer period Cepheids are subject to more reddening (presumably circumstellar) than the shorter period Cepheids. A similar effect has been found by Feast (1974) and he suggests that if this effect is taken into account there is no need for the composite P-L relationship derived by Sandage and Tammann (1968) to flatten out the long period end.

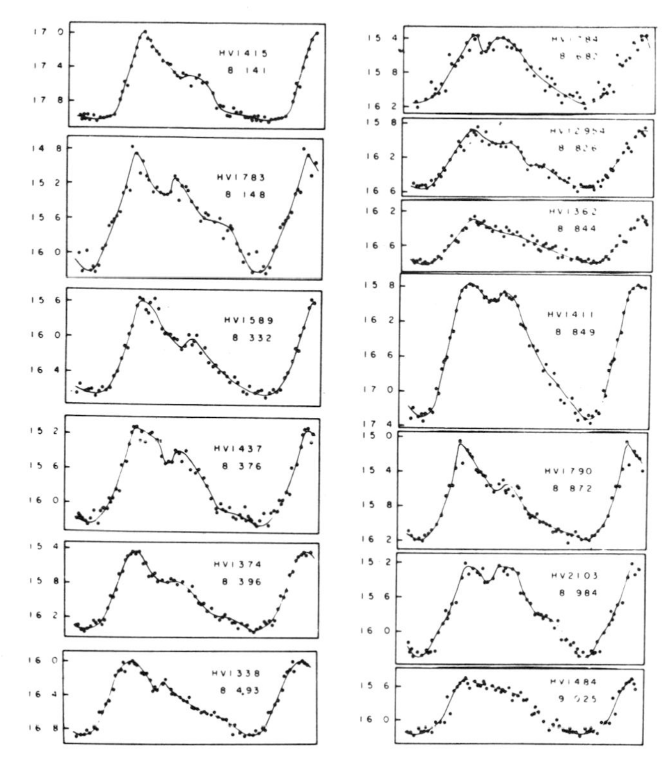

Figure 4. A selection of light curves showing conspicuous bumps amongst Population I Cepheids in the Small Magellanic Cloud (Payne-Gaposchkin and Gaposchkin 1966).

The systematics of Population I Cepheid light curve shapes have been known for a long time. The famous Hertzsprung progression of light curve shape with period may be described briefly as follows. The secondary bump in Cepheid light curves first becomes visible at periods near 7 days on the descending branch of the light curve. As the period is increased the phase of the bump decreases until it coincides with maximum light for periods near 10 days. For periods greater than 10 days the bump generally appears on the ascending branch of the light curve until for periods greater than about 15 days the bump disappears. A subsection of the Hertzsprung progression is illustrated in Fig. 4. This progression has been found in all galactic systems where Population I Cepheids have been observed and the relationship between the phase of the bump and the period is apparently identical from one galaxy to another (van Genderen 1970). The Hertzsprung progression has been found in the non-linear non-adiabatic models of pulsating stars and the phase of the bump varies systematically with period in exactly the manner predicted by the observations (Christy 1968, Stobie 1969a, b). The phase of this bump together with the period of a variable star provides a potential method of deriving the mass and radius of the star (Fricke et al. 1972). Masses and radii derived in this way, however, are systematically too low compared with masses derived from evolutionary theory and there is some uncertainty at present attached to the theoretical calibration of this relation (Stobie 1974).

The light and radial velocity curves of Population I Cepheids are known to be almost mirror images of one another (Cox 1974). This

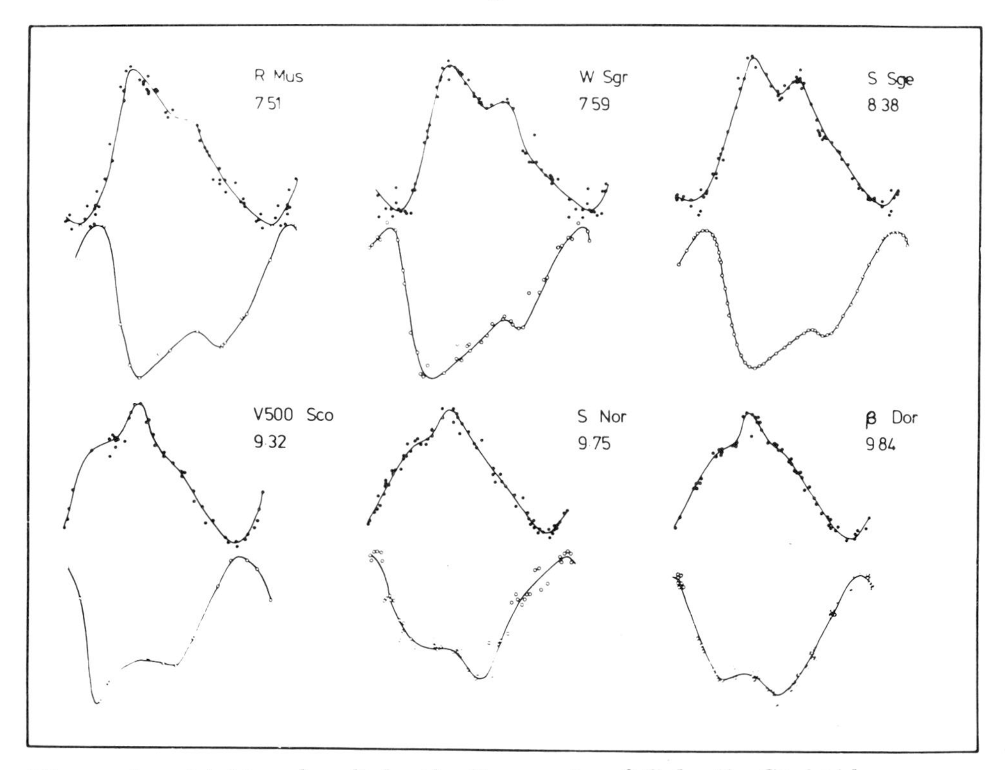

Figure 5. Light and radial velocity curves of Galactic Cepheids.

is not true, however, in the case of Cepheids with a secondary bump in their light curve (Fricke et al. 1972). A study of six Population I Cepheids with accurate light and radial velocity curves (Stobie, unpublished) has shown that there is a systematic delay of 1.26 days between the appearance of the bump on the light curve and the appearance of the bump on the velocity curve (Fig. 5). This delay is quantitatively what would be expected if the bump corresponds to a pulse travelling with the local speed of sound (Christy 1968) which first encounters the photosphere and subsequently reaches the line forming region of the atmosphere.

The question of the mode of a pulsating star is one of fundamental importance in any application of the P-L or P-L-C relationships. Unfortunately, unlike the RR Lyrae stars, the Population I Cepheids do not form a clear cut distinction in a period-amplitude plot into first overtone and fundamental mode pulsators. Theory predicts that the first overtone Cepheid pulsators will occur at the blue side and short period end of the instability strip (Stobie 1969b, King et al. 1973). Observational evidence in favour of this is the study by the Gaposchkins of Cepheids in the SMC (Payne-Gaposchkin and Gaposchkin 1966). They isolated a group of Cepheids with sinusoidal, low amplitude light curves (which we tentatively identify with first overtone pulsators). This group of Cepheids occurs in the period range $1 < P < 3$ days and systematically lies about $0^{m}.5$ above the mean P-L relation. Nikolov and Tsvetkov (1972) in a study of the pulsation amplitude of Galactic Cepheid variables isolated a group of low amplitude variables in a $\log \Delta R - \log P$ plot. Although Nikolov and Tsvetkov do not claim these variables are first overtone pulsators we would consider them likely candidates.

Theoretically the question as to which mode a star will pulsate in has been clarified recently although a complete physical understanding of the process is not at hand (Iben 1971, Cox et al. 1972b, Tuggle and Iben 1973 and Stellingwerf 1975a, b). Although the majority of these results has been obtained in applications to RR Lyrae variables we expect that similar results will be obtained in Cepheid variables. Van Albada and Baker (1973) proposed that in RR Lyrae stars there exists a finite transition region between fundamental and first overtone pulsators in which a star will persist pulsating in the mode with which it entered the transition region. Thus the pulsation mode with which stars are observed in this transition region depends upon their evolutionary history. Subsequent non-linear pulsation calculations studying the modal stability of models (Stellingwerf 1975a, b) have confirmed this prediction for RR Lyrae models and extended the result to low mass Cepheid models. The results (Fig. 6) show that a strip of width $\Delta T_e \sim 200$ - 300°K appears to exist in which a model can pulsate in either fundamental or first overtone pulsation.

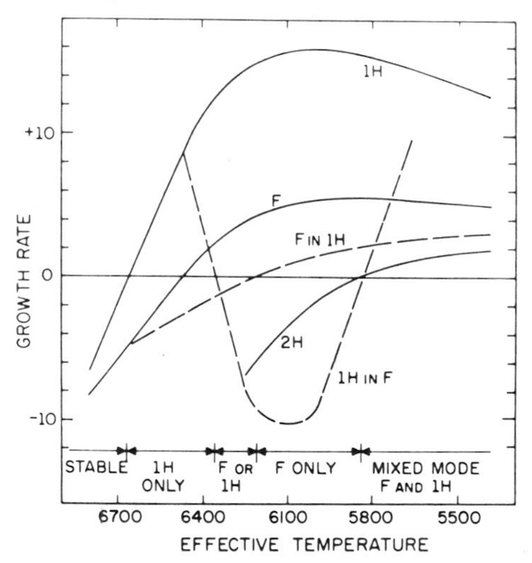

Figure 6. Composite growth rate diagram for models with $M/M_{\odot}$ = 0.6, P_0 = 2.6 days. Solid lines: linear growth rates; dashed lines: approximate non-linear switching rates ("F in 1 H" is switching rate of 1 H towards F). The various regions of different non-linear behaviour are indicated below the curves (Stellingwerf 1975b).

Apart from a subset of Cepheids known as double-mode Cepheids the majority of Population I Cepheids appear to have very stable light curves (Asteriadis <u>et al.</u> 1974). Exceptions to this rule are two long period Cepheids, ℓ Car (Feinstein and Muzzio 1969) and IU Cyg (Tammann 1969). Most Population I Cepheids also appear to have stable periods. The few Cepheids which have significant changes in their period show a general tendency in that the greater the period of the star the greater the period changes observed (Parenago 1958). These period changes have been further studied by Balazs-Detre and Detre (1965) and in no case can the period changes be assigned unambiguously to expected evolutionary changes in the star. Rather the (O-C) diagram is characteristic of random period fluctuations.

Recently evidence has been advanced for the existence of non-pulsating stars in the Population I instability strip (Fernie and Hube 1971, Schmidt 1972). It seems unlikely that the observational errors can be sufficiently large to exclude these stars from the instability strip and they consequently present a problem to the theoretician. Cox et al. (1973) have proposed that these stars may be deficient in helium, at least in the region of the helium ionization zones.

3. Population II Cepheids

The evolutionary status of Population II Cepheids has clarified considerably in the last ten years. Stars on the horizontal branch are believed to be in the stage of core helium burning. As the helium in the core is exhausted the stars evolve to higher luminosities and eventually approach the asymptotic red giant branch. Depending upon their initial position on the horizontal branch and helium content the stars can subsequently evolve through the instability strip in a supra-horizontal branch stage (Fig. 7). These suprahorizontal branch stars explain the existence of the short period group of Population II Cepheids in the period range 1 - 2 days observed in globular clusters and in the field. As the star ascends the asymptotic red giant branch it is subject to thermal pulses during the helium shell burning stage. These pulses cause it to execute loops in the H-R diagram which may intersect the instability strip (Schwarzschild and Härm 1970). This stage of evolution is thought to correspond to the longer period group of Population II

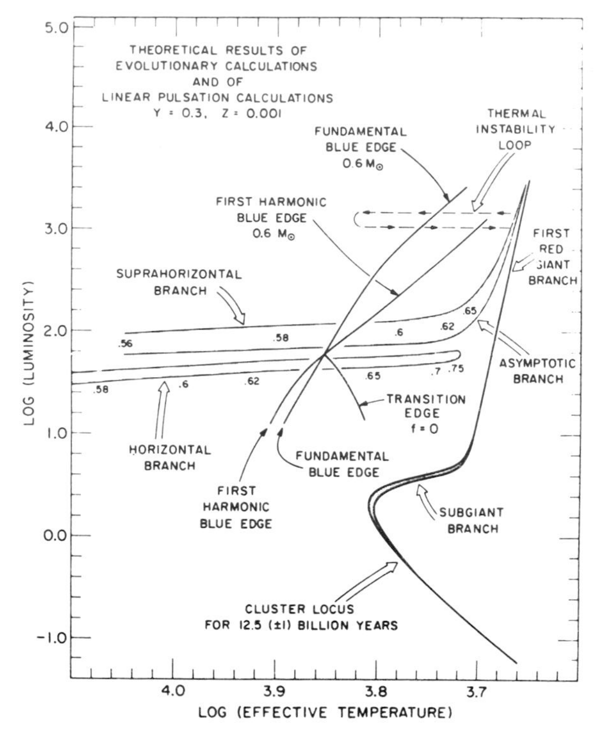

Figure 7. Theoretical H-R diagram showing results of evolution and linearized pulsation calculations for Population II stars having Y = 0.3, Z = 0.001. The numbers along the horizontal and suprahorizontal branches give the stellar masses, in solar units, and the mean locations of stars of the respective masses (Iben and Huchra 1971).

Cepheids in the period range 8 - 25 days.

Masses of Population II Cepheids have been derived from the P-R-M relationship (Böhm-Vitense et al. 1974). The results give $M/M_\odot \sim 0.55$ consistent with the pulsation masses derived for RR Lyrae stars. These masses imply that the stars have at some stage (presumably before the horizontal branch phase) lost mass as their main sequence progenitors have masses nearer $M/M_\odot \sim 0.8$ (Iben 1972).

The helium content of Population II Cepheids can be determined from the agreement of the observed and theoretical blue edges.

Demers and Harris (1974) find that a value of Y ~ 0.3 is predicted by their observations. This is consistent with the helium content predicted for RR Lyrae variables (Christy 1966b) and also with the helium content of globular cluster stars determined by the ratio of the number of horizontal branch stars to the number of red giant stars with magnitude brighter than the mean magnitude of the horizontal branch (Iben 1972).

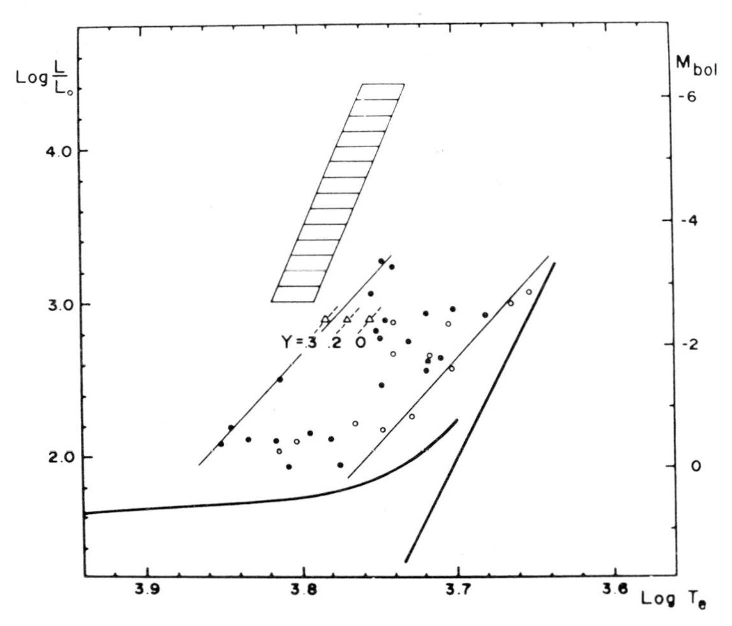

Figure 8. H-R diagram for Population II Cepheids. Hatched region is Population I instability strip. The theoretical blue edges for three different helium contents are shown (Demers and Harris 1974).

The location of Population II Cepheids in the H-R diagram has been studied by Demers and Harris (1974). They find that the Population II Cepheid instability strip is considerably broader and flatter than the Population I Cepheid instability strip (Fig. 8). P-L relationships for Population II Cepheids have been derived by a number of authors (Fernie 1964, Kwee 1968b and Demers and Wehlau 1971). The slope of the relationship is shallower than the slope of the corresponding Population I Cepheid relationship. This difference is a consequence of the Population II variables all having a similar mass independent of their luminosity. Caputo and Castellani (1972) have investigated P-L, P-C relationships and amplitude dependences for globular cluster Cepheids. They find correlations between the amplitude, colour and luminosity of a variable in the strip in the sense that for $\log P < 1$ largest amplitude occurs for the brightest and bluest variables whereas for $\log P > 1$ the correlations are reversed.

Relatively few pulsation calculations have been carried out relevant to Population II Cepheids. Some linear, non-adiabatic calculations have been reported by Cox and King (1972) and by Cox et al. (1973). Non-linear, non-adiabatic models of W Virginis have been constructed by Christy (1966c) and Davis (1972). The model by Davis showed a stillstand on the descending branch of the light curve which he attributed to the formation of a shock at the photosphere.

The light curves of field Population II Cepheids have been studied by Kwee (1968a). He finds that bumps or shoulders are a common feature in the light curves. In particular in the group of longer period Cepheids, $13 < P < 19$ days, a shoulder on the descending branch of the light curve is a frequent occurrence (cf stillstand in theoretical light curve of Davis 1972). The short period group of Cepheids, $1 < P < 3$ days, also show conspicuous bumps in their light curves. Theoretically bumps at this period are expected from the same mechanism that causes bumps in the Population I Cepheid phenomenon (Christy 1970). It has been proposed that a Hertzsprung progression amongst Population II Cepheids may exist similar to that observed in Population I Cepheids and observational evidence has been presented in support of this (Stobie 1973, Mandel 1971).

There is little data on the stability of the light curves of Population II Cepheids from epoch to epoch. One extreme example of an unstable light curve is RU Cam (Broglia and Guerrero 1973). In general Population II Cepheids appear to be much more subject to abrupt period changes than Population I Cepheids, the frequency and amplitude of the period changes being an order of magnitude greater (Parenago

1958, Balazs-Detre and Detre 1965 and Kwee 1967).

4. Double-mode Cepheids

Oosterhoff (1957a) was first to recognize the existence of a distinct class of Cepheid variable characterized by an abnormally large scatter in their photoelectric light curves. Subsequent analysis of the photoelectric observations revealed that the observations could be satisfactorily accounted for by the light curve being modulated with beat period, P_b. Interpreting this variation as being caused by the superposition of two component periodicities of period P_0, P_1 we have the relation

$$\frac{1}{P_1} - \frac{1}{P_0} = \frac{1}{P_b} \quad . \tag{1}$$

Note that there is an ambiguity in equation (1) in attempting to determine the secondary period if the only information available is the primary period and the beat period. The value of secondary period depends upon whether the primary period is assigned to P_0 or to P_1. This ambiguity, however, can be resolved if, after the primary period has been found, the mean curve corresponding to this primary period is subtracted from the data and the residuals then searched for the second period. The presence of two periodicities has also been found in the radial velocity observations (Oosterhoff 1957b).

The number of double-mode Cepheids whose light curves have been analysed for their component periodicities totals eight. The light curve characteristics of seven of these are listed in Table 1. The one Cepheid omitted is V439 Oph (Gusev 1967) on the basis that photoelectric observations by Sturch (1966) show no scatter in excess of observational error. The primary periods of the Cepheids lie in the range 2 to 4 days and the period ratios in the narrow range 0.703 $P_1/P_0 < 0.711$. The amplitudes ΔV, ΔB of the two periodicities have been calculated according to the method of Stobie (1970). It is apparent that in all cases (except AX Vel) the amplitude of the longer period exceeds the amplitude of the shorter period.

Table 1

Periods and amplitudes of double-mode Cepheids

Star	P_0	P_1	P_1/P_0	ΔV_0	ΔV_1	ΔB_0	ΔB_1	Reference
Y Car	3.6398	2.5590	0.7031	0.58	0.29	0.78	0.36	Stobie 1972
TU Cas	2.1393	1.5183	0.7097	0.66	0.31	0.99	0.46	Oosterhoff 1957b
BK Cen	3.1739	2.2366	0.7047	0.52	0.20	0.81	0.28	Leotta-Janin 1967
VX Pup	3.0117	2.136	0.709	0.46	0.33	0.52	0.49	Stobie 1970
U TrA	2.5684	1.8249	0.7105	0.47	0.25	0.79	0.35	Oosterhoff 1957a, Jansen 1962
AP Vel	3.1278	2.1993	0.7031	0.55	0.41	0.79	0.60	Oosterhoff 1964
AX Vel	3.6731	2.5928	0.7059	0.22	0.33	0.28	0.49	Stobie and Hawarden 1972

The two periods present P_0, P_1 have been identified with the periods of the fundamental and first overtone radial modes of a pulsating star. This is primarily because the period ratio corresponds very closely to the value expected theoretically. Note that the values of P_0, P_1 derived for VX Pup in Table 1 are uncertain. Fourier analysis (Stobie, unpublished) of the observations of VX Pup (Mitchell et al. 1964, Takase 1969) has shown that there are a number of possible solutions to the periods present in this star. Further photoelectric observations are required before a definitive solution can be given.

Unlike the case of some delta Scuti stars where the periods present at one epoch of observation may not predict the behaviour of the star at another epoch it appears that the periods present in a double-mode Cepheid are stable (e.g. results of Jansen 1962). Although the superposition of the two mean curves corresponding to the periods P_0, P_1 explains the main characteristics of the light variation, it does not precisely reproduce the observed light curve because non-linearities are present in the interaction of one pulsation mode with another. Non-linearities in the superposition of the modes are clearly seen in the observations of U TrA (Oosterhoff 1957a) since the amplitude variation at maximum light is greater than the amplitude variation at minimum light.

Table 2 gives a list of known and suspected double-mode Cepheids. Because of the observed period range of known double-mode Cepheids (from 2 to 4 days) we have restricted the search for further candidates to the period range 1 to 5 days. Only Cepheids with photoelectric observations have been considered. The data are taken from the catalogue of Schaltenbrand and Tammann (1971). The standard deviations σ_B, σ_V show that these stars have a much larger scatter about their mean light curve than would be expected for a single-mode Cepheid. There are a number of reasons as to why a Cepheid might have a spuriously large scatter in its light curve (wrong period, variable period, systematic differences between photometric systems of observers). In drawing up the list of suspected double-mode Cepheids we have attempted as far as possible to exclude variables subject to the

Table 2

Data on known and suspected double-mode Cepheids
in the period range 1 < P < 5 days

Star	P	<V>	<B-V>	<U-B>	σ_B	σ_V
Y Car	3.64	8.13	0.64	0.38	0.186	0.136
DY Car	4.67	11.36	1.08	-	0.074	0.050
EY Car	2.88	10.34	0.85	0.58	0.072	0.065
FZ Car(1)	3.58	11.85	1.18	0.82	-	-
GZ Car	4.16	10.22	1.00	-	0.067	0.051
TU Cas	2.14	7.72	0.61	0.39	0.131	0.094
UZ Cen	3.33	8.77	0.79	0.52	0.082	0.064
BK Cen	3.17	9.99	0.90	0.61	0.108	0.071
TZ Mus	4.94	11.66	1.35	0.99	0.144	0.180
VX Pup	3.01	8.18	0.61	0.42	0.140	0.089
U TrA	2.57	7.93	0.64	0.39	0.100	0.071
AP Vel	3.13	10.03	1.07	-	0.141	0.098
AX Vel	2.59	8.23	0.72	0.48	0.089	0.071

Note: (1) This star is not in Schaltenbrand and Tammann's (1971) catalogue because the observations are poorly distributed in phase. The scatter, however, is still considerable and the star is included as requiring further observation.

above spurious effects. It should be noted that there are a number of variables mentioned in Kukarkin et al. (1969) as showing variations in their light curve but these have not been included in Table 2 unless the photoelectric observations also showed an unambiguous scatter.

The variables listed in Table 2 form a significant fraction of the short period Cepheids with photoelectric observations in Schaltenbrand and Tammann's catalogue as is shown in Table 3.

Table 3

Fraction of double-mode Cepheids as a function of period

Period range (days)	Number of Cepheids			Fraction of double-mode
	type I	type II	double-mode	
1.0 - 2.0	2	9	0	0.00
2.0 - 3.0	3	3	4	0.40
3.0 - 4.0	20	4	6	0.20
4.0 - 5.0	42	2	3	0.07

Although there may be some selection effects at work there is no strong reason to suppose that these numbers are not representative of the relative fraction of double-mode Cepheids in the solar neighbourhood. Thus the double-mode Cepheids form a significant subset of solar neighbourhood Cepheids in the period interval 2 to 4 days.

All the variables in Table 2 have been classified as type I Cepheids although TU Cas has been classified both as a type I (Mianes 1963) and a type II Cepheid (Petit 1968). None of the variables lies more than 250 parsecs from the galactic plane (assuming luminosities consistent with Population I Cepheids). Thus the properties of double-mode Cepheids are consistent with them being members of the young disc or Population I Cepheids.

In view of the relative fraction of double-mode Cepheids in the solar neighbourhood it is at first sight somewhat surprising that these variables have not been found in external galaxies (especially the SMC where Cepheids in the period range 2 to 4 days are abundant). The scatter in the light curves of some known double-mode Cepheids is sufficiently high (a range of $0^{m}.5$ at maximum light) that there should be no difficulty detecting these stars photographically if they exist. Wesselink and Shuttleworth (1965) comment on a few SMC Cepheids with periods near 3 days which exhibited considerable irregularities in their light curve. Stobie (unpublished) has Fourier analysed the observations by Butler (1975a, b) of 20 Cepheids in the LMC and SMC for which it had proved difficult to determine the primary period present. In no case was there any evidence that the difficulty in determining the period or the existence of a large scatter in the light curve was caused by the presence of two periods.

The importance of the double-mode Cepheids lies in the fact that with two periods present, which we identify with the fundamental and first overtone modes of radial pulsation, there is the possibility of deriving the masses and radii of the stars solely from a knowledge of these two periods (Petersen 1973). The ratio P_1/P_0 also determines the pulsation constant Q_0 and hence the mean density of the star, giving values which are almost model independent (Fitch 1970). It is well known that the period of a variable star is determined principally by its radius and mass. The period of the linear, adiabatic model also gives values very close to the period derived from the non-linear, non-adiabatic model. The recent calculations by Stellingwerf (1975b) have shown that in a theoretical model of a double-mode Cepheid the interaction of one mode with another does not change the periods derived from the single-mode calculations by more than 0.5 percent. Thus if we know the theoretical P_0-R-M and P_1-R-M relationships we can in principle calculate R and M. The results (Fig. 9) show that the masses and radii of double-mode Cepheids lie in the range $0.7 < M/M_\odot < 1.7$, $14 < R/R_\odot < 23$ (Petersen 1973, Stellingwerf 1975b).

The evolutionary stage of the double-mode Cepheids is not clear at present. The masses and radii derived from the periods P_0, P_1 are inconsistent with the masses and radii expected for Population I Cepheids of similar period (i.e., $M/M_\odot \sim 4.5$, $R/R_\odot \sim 30$). Accepting the low radii of these double-mode Cepheids together with their effective temperature ($T_e \sim 6200^\circ$K) leads to a luminosity of $M_{bol} \sim -2$, which is approximately $0^m.7$ fainter than the absolute magnitude of Population I Cepheids of similar period. The spectroscopic gravities of these variables give conflicting results as Rodgers and Gingold (1973) analysing U TrA found that it had a gravity consistent with the gravity of

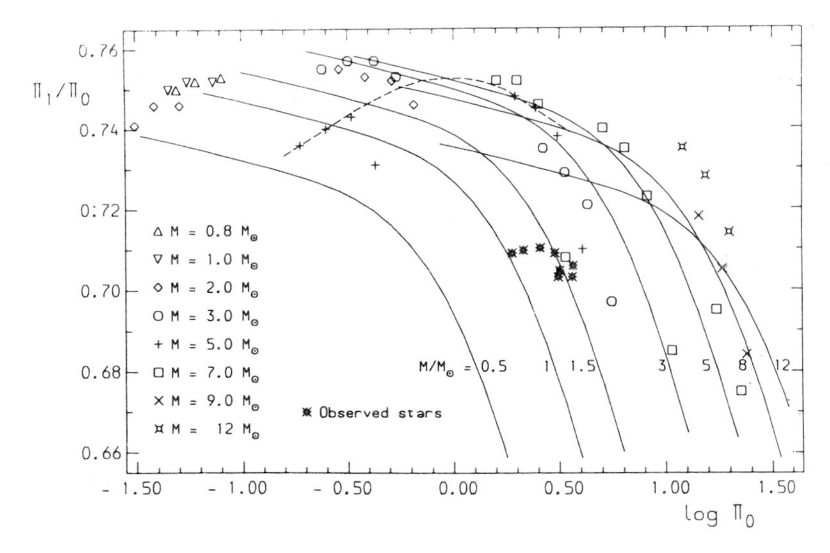

Figure 9. Period ratio as function of the period of the fundamental mode for an extreme Population I composition. Full curves are based on fitting formulae given by Cox et al. (1972a); the number at each curve gives the mass in solar units (Petersen 1973).

Population I Cepheids of similar period. On the other hand Schmidt (1974) finds that TU Cas has a higher gravity than a Population I Cepheid.

Stellingwerf (1975a, b) has presented new and exciting results on the modal behaviour of pulsating stars (Fig. 6). From an examination of the stability of the fundamental mode and first overtone limit cycle pulsations there seem to be two possibilities for the occurrence of double-mode Cepheids. One is that they have been caught in the stage of switching from one mode to another mode (F or 1 H region in Fig. 6). This was the supposition made previously (Stobie 1970, Rodgers 1970) that these stars were in a transition region between fundamental and first overtone pulsators. Stellingwerf's results, however, indicate that the timescale for switching modes is so short that there is little chance of observing a star in this stage. Instead it seems more likely that these double-mode Cepheids occur at the red edge of the instability strip where a stable mixed-mode behaviour can exist (Fig. 6). Mixed-mode models occur for temperatures, $T_e < 5850^oK$ which may be compared with the observed effective temperatures of $T_e \sim 6200^oK$ for double-mode Cepheids (Schmidt 1972, Rodgers and Gingold 1973).

It has been proposed (Petersen 1973, Stellingwerf 1975b) that the double-mode Cepheids represent a new type of variable star occurring in the instability strip intermediate in luminosity and mass between the RR Lyrae variables and the Population I Cepheids. The difficulty with this hypothesis, however, is that no evolutionary models of this mass have been found to enter the instability strip during the core helium burning stage.

We consider that it is not excluded that the double-mode Cepheids are in fact normal Population I Cepheids lying at the red edge of the instability strip. This explanation is tantamount to saying that the masses and radii derived from the P-R-M relationships are wrong. From the theoretical point of view the period of a pulsating star is one of the most accurately determined parameters in the calculations of models of pulsating stars. However, since the powers of R and M in the P_0-R-M and P_1-R-M relationships (in the region relevant to double-mode Cepheids) are similar, the calculation of R or M becomes very sensitive to small changes in the constants in these relationships (Stobie 1974). It may be that some change in the input physics to the model calculations could produce masses and radii consistent with Population I Cepheids. Petersen (1974), however, has investigated some arbitrary changes in the opacity and finds that these changes make little difference to the P-R-M relationships.

References

Albada, T. S. van and Baker, N. H., 1973. Ap. J., 185, 477.

Asteriadis, G., Mavridis, L. N. and Tsioumis, A., 1974. Stars and the Milky Way System, Proc. of First European Astronomical Meeting, Vol. 2, 17.

Baker, N. and Kippenhahn, R., 1965. Ap. J., 142, 868.

Balazs-Detre, J. and Detre, L., 1965. The Position of Variable Stars in the Hertzsprung-Russell Diagram, 3rd IAU Colloquium on Variable Stars, Bamberg, p. 184.

Bell, R. A. and Parsons, S. B., 1972. Ap. Letters, 12, 5.

Böhm-Vitense, E., Szkody, P., Wallerstein, G. and Iben, I., Jr., 1974. Ap. J., 194, 125.

Broglia, P. and Guerrero, G., 1973. Mem. Soc. Astron. Italiana, Nuova Ser., 44, 157.

Butler, C. J., 1975a. Astr. and Ap. Suppl., in press.

Butler, C. J., 1975b. Preprint

Butler, C. J., 1975c. Paper presented at Tercentenary Symposium, Royal Greenwich Observatory.

Caputo, F. and Castellani, V., 1972. Ap. Sp. Sc., 19, 423.

Christy, R. F., 1966a. Annual Rev. of Astr. and Astrophysics, 4, 353.

Christy, R. F. 1966b. Ap. J., 144, 108.

Christy, R. F. 1966c. Ap. J., 145, 337.

Christy, R. F. 1968. Q. J. R. astr. Soc., 9, 13.

Christy, R. F. 1970. J. R. astr. Soc. Canada, 64, 8.

Cogan, B. C., 1970. Ap. J., 162, 139.

Cox, A. N., King, D. S. and Tabor, J. E., 1973. Ap. J., 184, 201.

Cox, J. P., 1974. Rep. Prog. Phys., 37, 356.

Cox, J. P. and King, D. S., 1972. Dudley Obs. Reports No. 4, 103.

Cox, J. P., King, D. S. and Stellingwerf, R. F., 1972a. Ap. J., 171, 93.

Cox, J. P., Castor, J. I. and King, D. S., 1972b. Ap. J., 172, 423.

Davis, C. G., 1972. Ap. J., 172, 419.

Demers, S. and Wehlau, A., 1971. A. J., 76, 916.

Demers, S. and Harris, W. E., 1974. A. J., 79, 627.

Feast, M. W., 1974. M. N. R. A. S., 169, 273.

Feinstein, A. and Muzzio, J. C., 1969. Astr. and Ap., 3, 388.

Fernie, J. D., 1964. A. J., 69, 258.

Fernie, J. D. and Hube, J. O., 1971. Ap. J., 168, 437.

Fitch, W. S., 1970. Ap. J., 161, 669.

Fricke, K., Stobie, R.S. and Strittmatter, P.A., 1972. Ap.J., 171, 593.

Gascoigne, S.C.B., 1969. M.N.R.A.S., 146, 1.

Gascoigne, S.C.B. M.N.R.A.S., 166, 25P., 1974.

Genderen, A.M. van, 1969. B.A.N. Suppl., 3, 221.

Genderen. A.M. van Astr. and Ap., 7, 244. , 1970.

Gusev, E.B., 1967. Inf. Bull. Var. Stars, No. 186.

Hofmeister, E., 1967. Zs.f.Ap., 65, 194.

Iben, I., Jr., 1966. Ap.J., 143, 483.

Iben, I, Jr. 1967. Annual Rev. of Astr. and Astrophysics, 5, 571.

Iben, I, Jr. 1971. Ap.J., 168, 225.

Iben, I, Jr. 1972. Dudley Obs. Reports No. 4, 1.

Iben, I., Jr. and Huchra, J.P., 1971. Astr. and Ap., 14, 293.

Iben, I., Jr. and Tuggle, R.S., 1972, Ap.J., 173, 135.

Iben, I, Jr. 1975. Preprint.

Jansen, A.G., 1962. B.A.N., 16, 141.

Kelsall, T., 1972. Goddard Space Flight Centre Preprint X-641-72-365.

King, D.S., Cox, J.P., Eilers, D.D. and Davey, W.R., 1973. Ap.J., 182, 859.

Kukarkin, B.V., Kholopov, P.N., Efremov, Yu.N., Kukarkina, N.P., Kurochkin, N.E., Medvedeva, G.I., Perova, N.B., Fedorovich, V.P. and Frolov, M.S., 1969. General Catalogue of Variable Stars, 3rd edition, Moscow.

Kwee, K.K., 1967. B.A.N. Suppl., 2, 97.

Kwee, K.K., 1968a. B.A.N., 19, 260.

Kwee, K.K., 1968b. B.A.N., 19, 374.

Lauterborn, D., Refsdal, S. and Weigert, A., 1971. Astr. and Ap., 10, 97.

Ledoux, P. and Walraven, Th., 1958. Handbuch der Physik, Vol. 51, S. Flügge, ed. (Springer-Verlag), p. 353.

Leotta-Janin, C., 1967. B.A.N., 19, 169.

Madore, B.F., 1975a. Ap.J. Suppl. No. 285.

Madore, B.F 1975b. Paper presented at Tercentenary Symposium, Royal Greenwich Observatory.

Mandel, O. E., 1971. Perem. Zvezdy, 17, 347.

Mianes, P., 1963. Ann. Astrophys., 26, 1.

Mitchell, R. I., Iriarte, B., Steinmetz, D. and Johnson, H. L., 1964. Ton. y. Tac. Bol., 3, 153.

Nikolov, N. and Tsvetkov, Ts., 1972. Ap. Sp. Sc., 16, 445.

Oosterhoff, P. Th., 1957a. B. A. N., 13, 317.

Oosterhoff, P. Th., B. A. N., 13, 320. 1957b.

Oosterhoff, P. Th., B. A. N., 17, 448., 1964.

Parenago, P. P., 1958. Perem. Zvezdy, 11, 236.

Payne-Gaposchkin, C. and Gaposchkin, S., 1938. Variable Stars, Harvard Observatory Monographs No. 5.

" 1966, Smithsonian Contr. to Astrophys., 9, 1.

Petersen, J. O., 1973. Astr. and Ap., 27, 89.

Petersen, J. O., Astr. and Ap., 34, 309., 1974.

Petit, M., 1960. Ann. Astrophys., 23, 681.

Robertson, J. W., 1971. Ap. J., 170, 353.

Rodgers, A. W., 1970. M. N. R. A. S., 151, 133.

Rodgers, A. W. and Gingold, R. A., 1973. M. N. R. A. S., 161, 23.

Sandage, A. and Tammann, G. A., 1968. Ap. J., 151, 531.

" 1969. Ap. J., 157, 683.

" 1971. Ap. J., 167, 293.

Schaltenbrand, R. and Tammann, G., 1971. Astr. and Ap. Suppl., 4, 265.

Schmidt, E. G., 1972. Ap. J., 174, 605.

Schmidt, E. G., Ap. J., 176, 165., 1972.

Schmidt, E. G., M. N. R. A. S., 167, 613., 1974.

Schwarzschild, M. and Härm, R., 1970. Ap. J., 160, 341.

Stellingwerf, R. F., 1975a. Ap. J., 195, 441.

Stellingwerf, R. F. Ap. J., 199, 705., 1975b.

Stobie, R. S., 1969a. M. N. R. A. S., 144, 485.

Stobie, R. S. 1969b. M. N. R. A. S., 144, 511.

Stobie, R. S. 1970. Observatory, 90, 20.

Stobie, R. S. 1972. M. N. R. A. S., 157, 167.

Stobie, R. S., 1973. Observatory, 93, 111.

Stobie R. S.,1974. Stellar Instability and Evolution, IAU Symposium No. 59, p. 49.

Stobie, R. S. and Hawarden, T., 1972. M. N. R. A. S., 157, 157.

Sturch, C., 1966. Publ. A. S. P., 78, 210.

Takase, B., 1969. Tokyo Astron. Bull. Second Series No. 191, p. 2233.

Tammann, G. A., 1969. Inf. Bull. Var. Stars No. 366.

Tuggle, R. S. and Iben, I., Jr., 1973. Ap. J., 186, 593.

Wesselink, A. J. and Shuttleworth, M., 1965. M. N. R. A. S., 130, 443.

Discussion to the paper of STOBIE

FITCH: Your earlier figures showed that 60% of all 2-4 day Cepheids are double mode variables. Does it seem reasonable that all of them should be at the red edge of the strip?

STOBIE: Yes, it just means that they don't penetrate so far into the strip from the red side.

PETERSEN: You showed a diagram from my paper on double mode Cepheids. The points representing individual models were taken from the investigation by Cogan in 1970. These results agree very well with the fitting formulae of Cox, King, and Stellingwerf from 1972. I tried to vary the chemical composition, and found that the mass values derived depend little on the composition parameters.

FITCH: You get better agreement with evolution theory if you don't use the CKS formulae. Fitch and Szeidel developed interpolation formulae by fitting to Cogan's models, and got better (i.e. higher) masses for the double mode Cepheids. However, these were still below evolutionary calculation expectations.

WAYMAN: Is it not crucial as to whether double mode Cepheids are found in the Magellanic Clouds?

STOBIE: I have looked hard amongst data for the Cloud Cepheids and have been surprised to be unable to definitely find any. They would be found relatively easily from photographic surveys whether higher or lower luminosities apply.

KWEE: Regarding galactic Population II Cepheids, you said that there exist very little data on this kind of object. Recently, I have observed 14 Population II Cepheids with periods between 1 and 3 days. The observations were made with the UBV photometer at La Silla in Chile. The resulting light curves will be published in due time and can be seen at this meeting.

Multiple Periodic Variable Stars
IAU Colloquium No.29, Budapest, 1975

CEPHEID PULSATION THEORY AND MULTIPERIODIC CEPHEID VARIABLES*

A. N. Cox

Theoretical Division
Los Alamos Scientific Laboratory, University of California
Los Alamos, New Mexico 87545, U.S.A.

J. P. Cox

Joint Institute for Laboratory Astrophysics, University of Colorado
Boulder, Colorado 80302, U.S.A.

In this review of the situation with regard to the multiperiodic Cepheid variables, our subject matter is divided into four parts. The first discusses general causes of pulsation of Cepheids and other variable stars, and their locations on the H-R diagram. For this section we draw upon the work during the past 10-15 years of J. P. Cox, Baker, Kippenhahn, A. N. Cox, King, Christy, Castor, Stobie, Stellingwerf, Davey, Iben, and Tuggle, mostly with the small amplitude linear nonadiabatic radial pulsation theory. In the second section we review the linear adiabatic and nonadiabatic theory calculation of radial pulsation periods and their application to the problem of masses of double-mode Cepheids. Contributions discussed are by Cogan, J. P. Cox, King, Stellingwerf, Petersen, Hansen, and Ross. Periodic solutions, and their stability, of the nonlinear radial pulsation equations for Cepheids and RR Lyrae stars are considered in the third section. This research has been done by Stellingwerf with previous development of methods by Baker and von Sengbusch and current work by A. N. Cox and Davey at Los Alamos. In the last section we give the latest results on nonlinear, nonperiodic, radial pulsations for Cepheids and RR Lyrae stars. This work has been done by Stellingwerf, King, A. N. Cox, J. P. Cox, and Davey.

References to the first three sections are given by Ledoux and

*Work partially performed under the auspices of the United States Energy Research and Development Administration.

Walraven (1958) and by J. P. Cox (1974).

The basic causes of pulsation in most common types of variable stars are now reasonably well understood. These stars include the classical Cepheids, the RR Lyrae variables, the W Virginis variables, and the dwarf Cepheids and Delta Scuti variables. These stars lie in the Hertzsprung-Russell diagram in a long, narrow, almost vertical region above and to the right of the main sequence--roughly, from a few to a few hundred thousand solar luminosities and from some 5000 K to 8000 K. This region shown in Fig. 1 is sometimes called the Cepheid instability strip, or simply the instability strip. Even the basic destabilizing mechanism of the red variables such as the Mira variables which have cooler surface temperatures than those in the instability strip is probably fairly well understood.

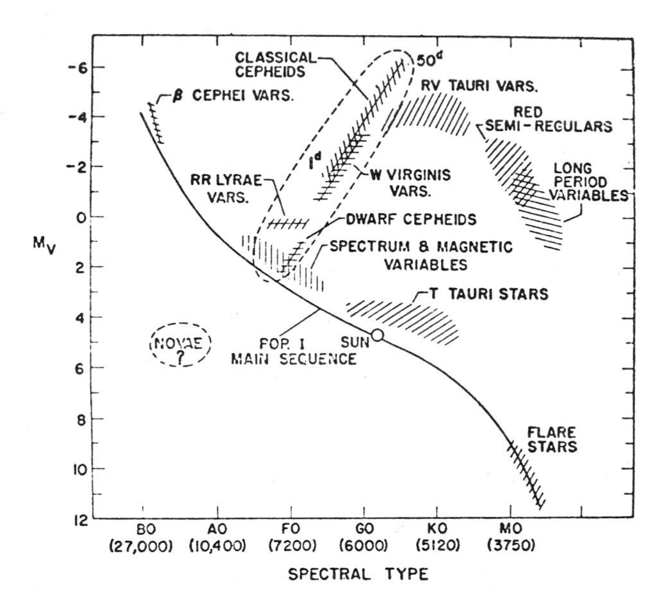

Fig. 1. The Hertzsprung-Russell diagram showing regions where variable stars are found. The double-mode Cepheids are found near the red edge of the Cepheid instability strip.

The instability mechanisms referred to above may be called envelope ionization mechanisms. They can produce pulsational instability because an abundant constituent of the star, such as hydrogen and/or helium, is in the midstages of ionization at a critical depth below the stellar photosphere. This partial ionization of a dominant constituent prevents the temperature in the regions of partial ionization from varying very much relative to the temperature in surrounding regions during the pulsation. This effect is called the gamma mechanism. In particular, the temperature in these regions does not increase very much during the compression phase of the pulsation cycle. This relative coolness during the compression phase, for example, has two consequences: first, it tends to impede the radiation flow out of these regions at this phase. Second, this partial ionization and relative coolness tend to cause the opacity of these regions to increase upon compression (the kappa mechanism), and thus, again, to impede the flow of radiation out of these regions at this phase. The net result of both of these factors is to prevent the escape of radiation, or to "trap" the radiation, during the compression phase. The trapped radiation then heats the relevant regions, and causes the pressure here to be somewhat larger, during the expansion phase than during the contraction phase. Hence, the pulsations tend to be pumped up, or excited, and thus to increase in amplitude with time, at least until nonlinear processes limit the amplitude to some finite value, presumably, to the observed values.

On the other hand, the deeper stellar layers, below the ionization regions, tend to behave in just the opposite manner and so tend to damp the pulsations. This effect of the deeper regions is sometimes called radiative damping.

Detailed calculations show that it is mainly the He^{+} - He^{++} -- the second helium ionization region -- which is responsible for the pulsations of stars in the instability strip, with a much smaller, though nonnegligible, contribution from hydrogen ionization. On the other hand, H ionization seems to be mainly responsible for the pulsations of the red variables.

Envelope ionization mechanisms have well accounted for most of the qualitative features of the above regions on the H-R diagram where many types of variable stars are found. For example, these mechanisms restrict most variable stars to relatively low effective temperatures. Also, the condition that the ionization regions lie at a great enough depth below the stellar surface implies that the hot, or blue, edges of instability regions should be fairly sharp and nearly vertical on an H-R diagram, as observed. This condition follows because this depth is considerably more sensitive to a star's effective temperature than to any other stellar parameter. The present general theories do not well account for the termination of instability on the red, or cool, edge of the instability strip. It is generally believed that this termination is caused by the onset of effective convection in the relevant parts of the stellar envelope, thus throttling the action of the above destabilizing mechanisms. However, the quantitative details of this throttling action still remain to be worked out.

In the linear theory of Cepheid pulsation with the destabilizing region near the surface, several modes of radial pulsation may be simultaneously excited. In most cases, observation and nonlinear theory show that only a single mode actually occurs at large, observable amplitudes. For these cases, or for the double-mode case which is occasionally also observed, what mechanism decides the actual behavior, and is there a chance to predict this behavior by the simpler, more efficient linear theory? For prediction of actual light and velocity curves, and the modal behavior, it seems that we must use nonlinear theory.

We now discuss linear theory results for periods.

It had been shown from the results of Cogan (1970) and of J. P. Cox, King, and Stellingwerf (1972) that the pulsation constant Q for a given mode depended mostly on the ratio M/R. These data from Cox, King and

Stellingwerf are in a plot as Fig. 2. Since Q is also proportional to the quantity $\sqrt{M/R^3}$ (M and R denote, respectively, stellar mass and radius) multiplied by the period, given a period and a radius (from an effective temperature and luminosity) a pulsation mass can be derived.

Petersen (1973) published an important paper in which he determined the masses and radii of eight double-mode Cepheids. He showed that period ratios are also related to M/R. Thus with two quantities, Π_0 and Π_1/Π_0 (Π = period, subscripts 0 and 1 denote, respectively, fundamental and first overtone), the double-mode fundamental period and the ratio of the two decomposed periods (presumed to be the fundamental and first overtone), the unknowns M and R can be found.

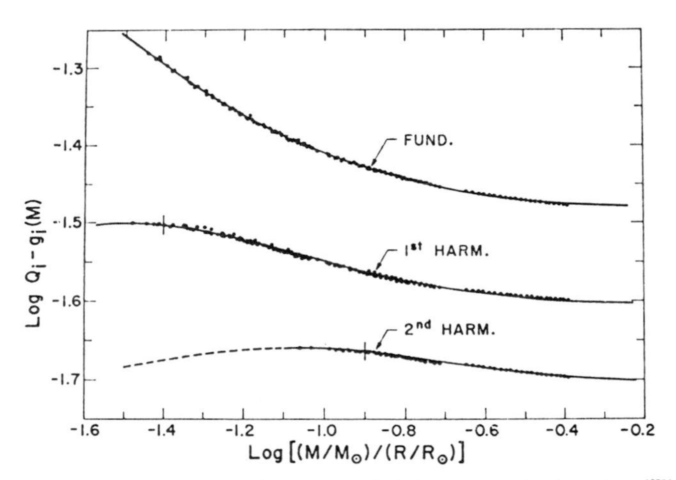

—Log $Q_i - g_i(\mathfrak{M})$ *versus* log $(\mathfrak{M}/R)$ for the Population I models, where the $g_i(\mathfrak{M})$ $(i = 0, 1, 2)$ are defined in eqs. (2)–(4), which are also plotted (*solid* and *dashed lines*). Solar units are used for $\mathfrak{M}$ and R, and the Q_i are in days. A few overlapping points have been omitted for clarity. No models lying to the left of the two vertical ticks are unstable in the 1H or 2H, respectively. The dashed part of the curve for the 2H indicates that this curve may not be a very accurate fit in this $\mathfrak{M}/R$ range.

Fig. 2. The pulsation constant Q for the lowest three radial pulsation modes as a function of M/R according to models of Cox, King, and Stellingwerf.

Results are shown in Fig. 3 which plots Π_1/Π_0 versus log Π_0 for various masses from linear theory data of Cox, King, and Stellingwerf and of Cogan. Observed stars are seen to have double-mode masses even lower than the possibly already low pulsation masses compared to evolutionary theory.

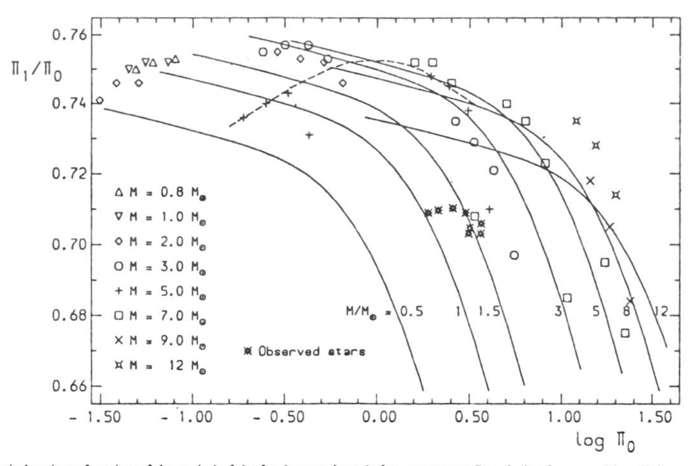

Period ratio as function of the period of the fundamental mode for an extreme Population I composition. Full curves are based upon the fitting formulae given by Cox *et al.* (1972); the number at each curve gives the mass in solar units. Individual points are taken from Cogan (1970), and the dashed curve is a roughly estimated mean curve for Cogan's models of 5 solar masses (see text). The variables given in Table 1 are also plotted in the figure

Fig. 3. The first overtone to fundamental period ratio as a function of fundamental period for various masses according to data collected by Petersen.

Further discussion by King, Hansen, Ross, and J. P. Cox (1975) showed that composition did not seem to affect greatly the Petersen argument. The same eight stars were redone with two more compositions and the results are given in Fig. 4. Periods in days are given in the second and third columns and masses and radii in solar units are given next for all three compositions -- the Kippenhahn (with hydrogen mass fraction X close to 0.6), King IVa, and King Va mixtures. Masses range from 1.04 to 2.16 solar masses -- not 3 to 5 as expected from evolutionary theory.

These authors then discuss UTrA in detail, assuming that its mass is 1.60 solar masses as given by the most reasonable population I composition and using the observed color. Results are in Fig. 5. For the four models considered, it appeared that luminosities like M_{bol} = - 1.4 to - 1.9 and T_e values approximately 5500 to 6100 K were needed to give the proper radius and simultaneous linear theory instability in both fundamental and first harmonic modes. These models are near the red edge of the instability strip.

MASSES AND RADII OF THE DOUBLE MODE CEPHEIDS

Star	Composition		CKS		X = 0.7		X = 0.8	
	Π_0	Π_1	M	R	M	R	M	R
V439 Oph	1.89	1.34	1.04	13.0	1.24	14.0	1.34	14.4
TU Cas	2.14	1.52	1.16	14.6	1.37	15.7	1.48	16.2
U TrA	2.57	1.83	1.13	17.3	1.60	18.6	1.74	19.2
VX Pup	3.01	2.14	1.50	19.8	1.76	21.2	1.92	21.9
AP Vel	3.13	2.20	1.40	19.6	1.64	21.0	1.81	21.8
BK Cen	3.17	2.24	1.46	20.1	1.70	21.5	1.88	22.4
Y Car	3.64	2.56	1.59	22.6	1.85	24.1	2.05	25.1
AX Vel	3.67	2.59	1.68	23.1	1.95	24.8	2.16	25.8

Fig. 4. Masses and radii of double-mode Cepheids for three compositions, Kippenhahn Ia, King IVa, and King Va, with, respectively X = 0.602, 0.7, and 0.8. Both equation of state and opacity are from tables computed at Los Alamos.

MODELS OF THE DOUBLE MODE CEPHEID U TrA

Mass	M_{bol}	T_{eff} (°K)	Π_1/Π_0	R	η_1/η_0
1.60	-1.4	5500	0.72	19.0	1
1.60	-1.6	5800	0.72	18.8	2
1.60	-1.9	6100	0.72	19.5	2
1.60	-2.2	6600	0.72	19.1	*

*For this model both fundamental and first harmonic modes were stable.

Fig. 5. Models of the double-mode Cepheid UTrA for 1.6 $M_{\odot}$ with the fundamental and first overtone periods held fixed at 2.57 and 1.83 days.

Moving now to periodic solutions of the nonlinear pulsation equations, we review briefly some results recently completed by Stellingwerf (1975). His method (1974) has been described as a Henyey method where the pulsation period and all starting values of temperature, radius, and velocity for Lagrangian mass zones at an arbitrary phase are iteratively adjusted so that conditions are exactly reproduced at the end of the true period. Stellingwerf has discussed mostly RR Lyrae stars, but he has also done work on two mixed-mode models for UTrA at 5800 and 6100 K--two models discussed by King, Hansen, Ross, and Cox.

Figure 6 gives the relevant results for the modal behavior of the two 1.6 $M_\odot$, 19 $R_\odot$, King IVa composition models. Obviously much more than the two models is needed for all the details of these curves which actually are based on Stellingwerf's (1975) RR Lyrae work also. Plotted are linear kinetic energy growth rates for the fundamental, first overtone and second overtone pulsations as solid lines. Blue edges are where growth rates become positive, indicating an increase in amplitude with time. A linear analysis of the stability of the nonlinear periodic pulsations shows that the first overtone will grow out of a pure fundamental pulsation between 6400 and 6500 K. At about 6500 K, the fundamental blue edge, the fundamental pulsation is just stable and so the linear growth of the first overtone from the fundamental is really the same as the growth of the first overtone from a static model in linear theory. No mode switching is predicted between about 6350 K and 6200 K, but at temperatures down to about 5800 K there at least is a tendency for the first overtone to switch to the fundamental. At and below about 5800 K mixed-mode behavior is expected.

Thus, proceeding from hot to cooler effective temperatures, there are four regions of distinct large-amplitude pulsation behavior: stable only (neither mode exists), first overtone only present, either mode occurring,

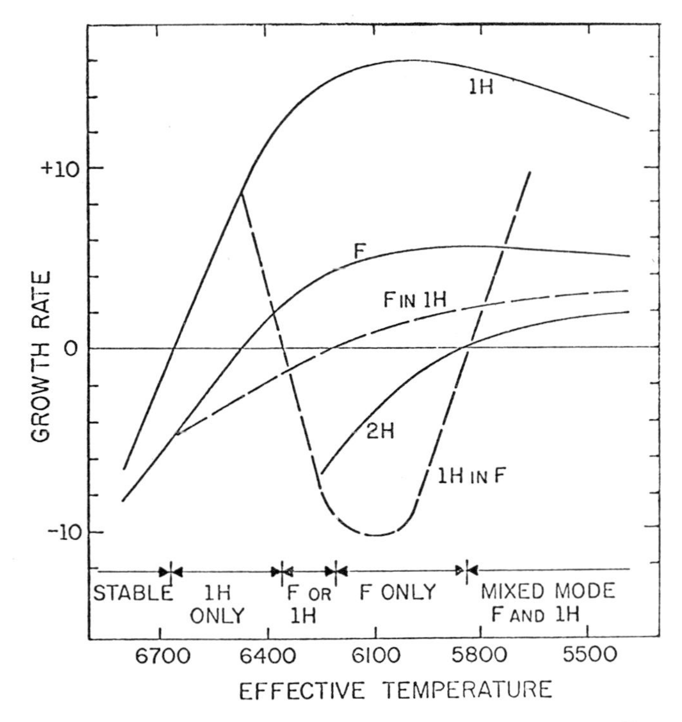

Fig. 6. Linear theory radial pulsation growth rates from static models and from nonlinear periodic solutions for the fundamental and first overtone modes. The Stellingwerf data are for UTrA models with 1.6 $M_{\odot}$ and 19 $R_{\odot}$ giving a constant period of 2.6 days.

fundamental only present, and then mixed fundamental and first overtone mode pulsations. There are no indications of switching from the fundamental or first overtone to the second overtone, which has a blue edge for this mass and luminosity at about 5800 K.

It is interesting to see what the stability analysis of the nonlinear pulsations gives for the RR Lyrae stars in the H-R diagram. This is shown in Fig. 7. The instability diagram just shown was for a line of constant radius, 19 $R_{\odot}$, and constant period, 2.6 days, in the H-R diagram. The same regions discussed before appear for these lower luminosities and lower masses. Note that the aperiodic or mixed mode region may lie entirely in the stable region to the red of the red edge. This may explain the very few (or zero) mixed mode RR Lyrae stars and in the case of higher masses the unlikelihood of finding multiperiodic Cepheids.

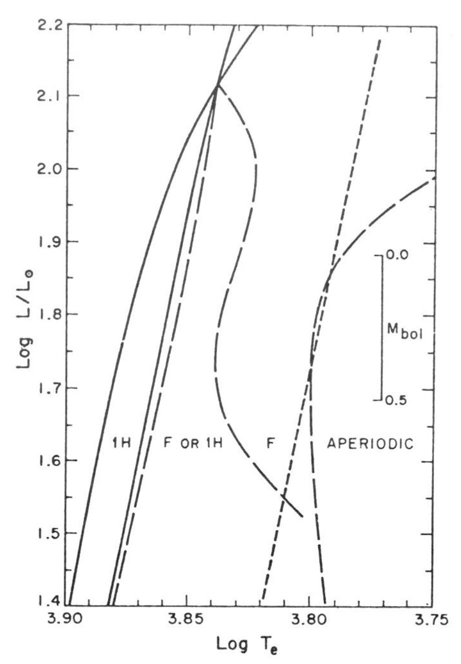

—H-R diagram showing the overall stability results for the main survey, $M = 0.578\ M_\odot$, $X = 0.7$, $Y = 0.299$, $Z = 0.001$. *Solid lines*, linear blue edges; *dashed lines*, nonlinear transition lines (type of behavior indicated); *dotted line*, estimated red edge.

Fig. 7. The instability strip in the region of the RR Lyrae stars on the theoretical Hertzsprung-Russell diagram. Modal behavior for 0.578 $M_\odot$ stars of the King Ia composition are indicated.

Figure 8 shows the only successful nonlinear, nonperiodic calculation, here actually for the case of the Stellingwerf RR Lyrae star model 2.6 with the fundamental period of 0.96 days and perhaps redward of the red edge. This case has been compared to the single observed mixed-mode RR Lyrae variable AC Andromedae with Π_0 = 0.711 day. His shorter period models do not show aperiodic behavior. Fitch, however, suggests a much larger mass for AC And. Thus there may be no double-mode low-mass RR Lyrae stars. We should point out that Stobie and King, J. P. Cox, Eilers, and Davey have mentioned mixed-mode behavior but these cases have not been well documented and may not be a permanent mixture of modes. It should also be pointed out that now at Los Alamos we are not able to reproduce this model 2.6 result with direct integrations through time.

King, A. N. Cox, Eilers, and J. P. Cox (1975) have attempted the direct initial value problem with integrations through time for the case of the mixed-mode Cepheid UTrA. As seen in Fig. 7, near the red edge, a star should show double-mode behavior. For $M_{bol} = -1.6$ and $T_e = 5600$ K, the complete time behavior calculated at Los Alamos is given in the next four figures. Figure 9 shows the growth of the pure first overtone mode to its limiting amplitude. No hint of any mixed mode appears even though a perturbation kinetic energy growth rate to the fundamental of several percent each period is expected according to Stellingwerf. The fundamental pulsations were followed in Fig. 10 with similar pure-mode behavior. This last result is greatly unexpected because, according to Stellingwerf's models and calculations, the linear growth rate from this fundamental mode toward the first overtone is over ten percent in kinetic energy per period. Also shown on this plot is a deliberate artificial perturbation, mixing the full amplitude fundamental velocities with three-quarters amplitude first overtone velocities. While there is mixed-mode behavior for a time, it all damps out to give a pure fundamental after about 100 periods. Figure 11 shows this artificial multiperiodic behavior in the velocity curve after six fundamental periods. Unfortunately after 250 periods the behavior given in Fig. 12 is indistinguishable from the pure fundamental before the perturbation was artificially imposed.

We note the following results to this date:

1. The linear theory gives periods and period ratios which can be interpreted in terms of very low masses for these observed stars, compared to evolutionary theory masses.

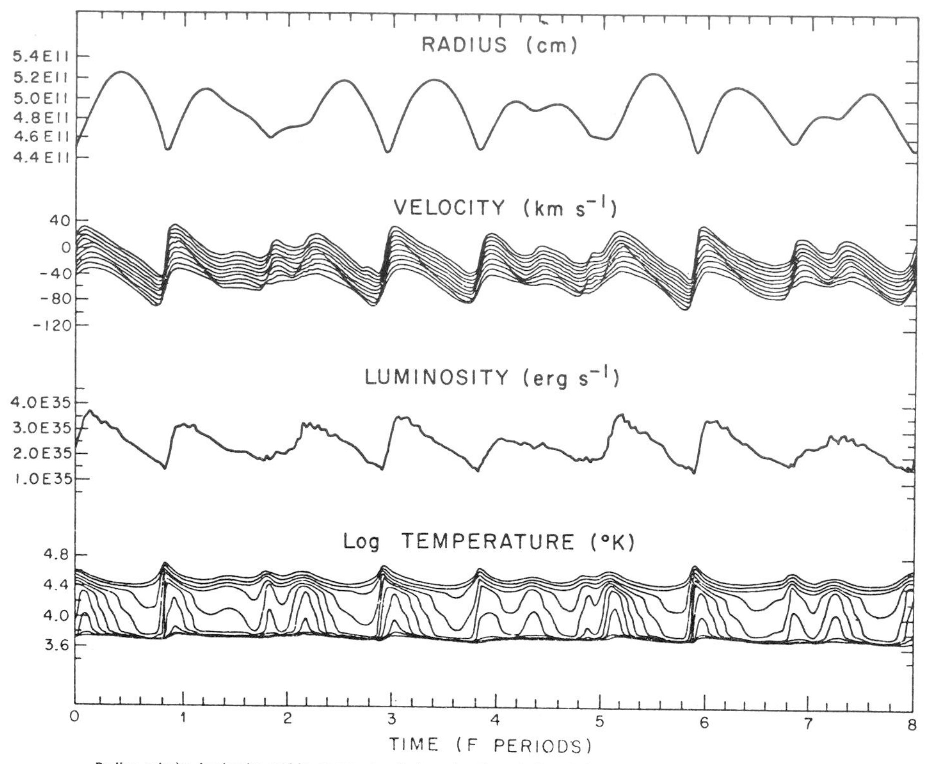

—Radius, velocity, luminosity, and log temperature (units and scales as indicated) for eight fundamental periods of the mixed-mode model described in the text. Successive velocity curves are shifted by 6 km s^{-1}.

Fig. 8. The time behavior of the Stellingwerf model 2.6. This calculation was made with an artificial advancing technique based on the stability analysis of the fundamental periodic limit cycle. The normal initial value technique gives identical results.

2. Certain nonlinear calculations predict only the permanent mixed-mode behavior for pulsators near the red edge of the instability strip. These calculations are based on a linear stability analysis of the full amplitude nonlinear periodic pulsation.

3. Direct initial value nonlinear calculations for RR Lyrae and Cepheid pulsations do not give any stable double-mode behavior except for the Stellingwerf case for his model 2.6 which cannot be confirmed by two distinct Los Alamos codes.

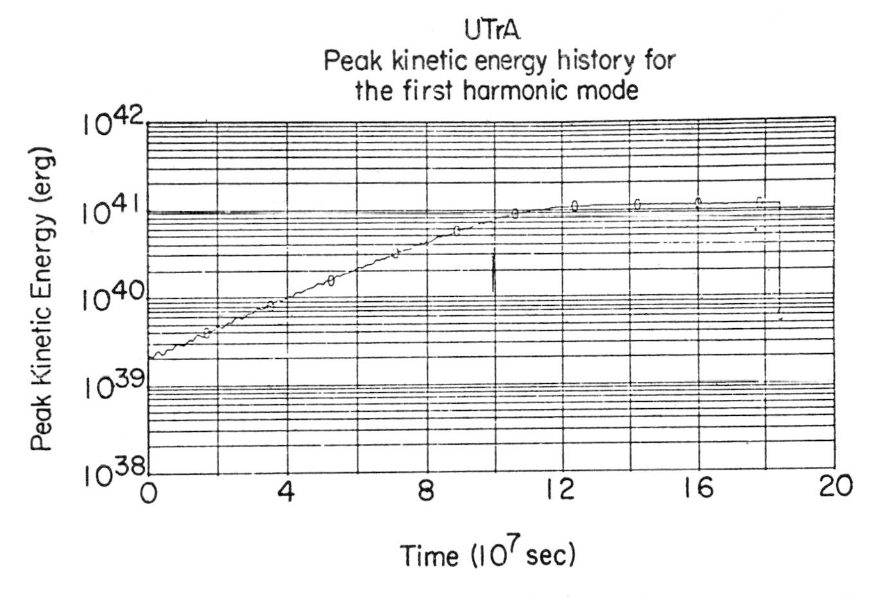

Fig. 9. The peak kinetic energy growth for the first overtone mode of a 1.6 $M_\odot$ UTrA model. An approximate velocity distribution is imposed on a hydrostatic structure.

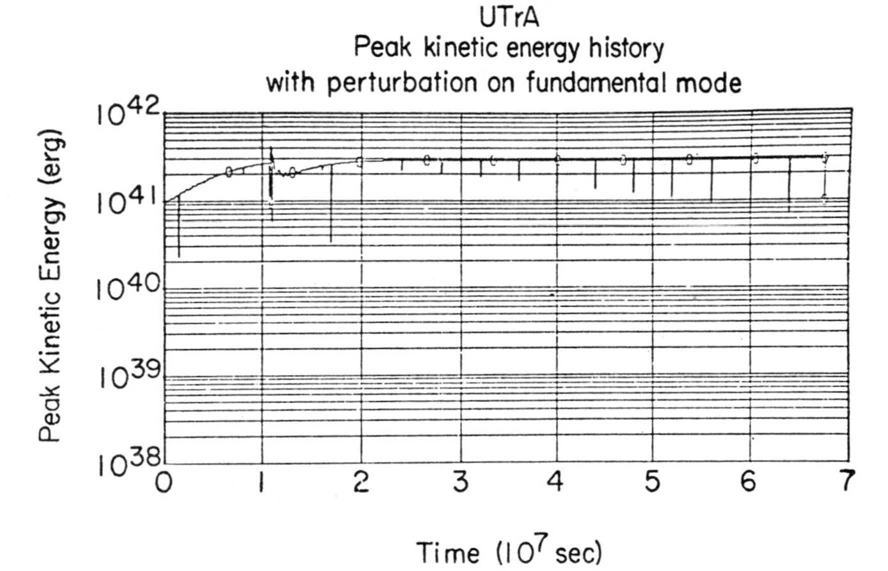

Fig. 10. The peak kinetic energy growth for the fundamental mode of a 1.6 $M_\odot$ UTrA model. After reaching the limiting level an artificial perturbation was introduced to induce double-mode behavior. After less than 50 periods the motion reverts to the pure fundamental mode.

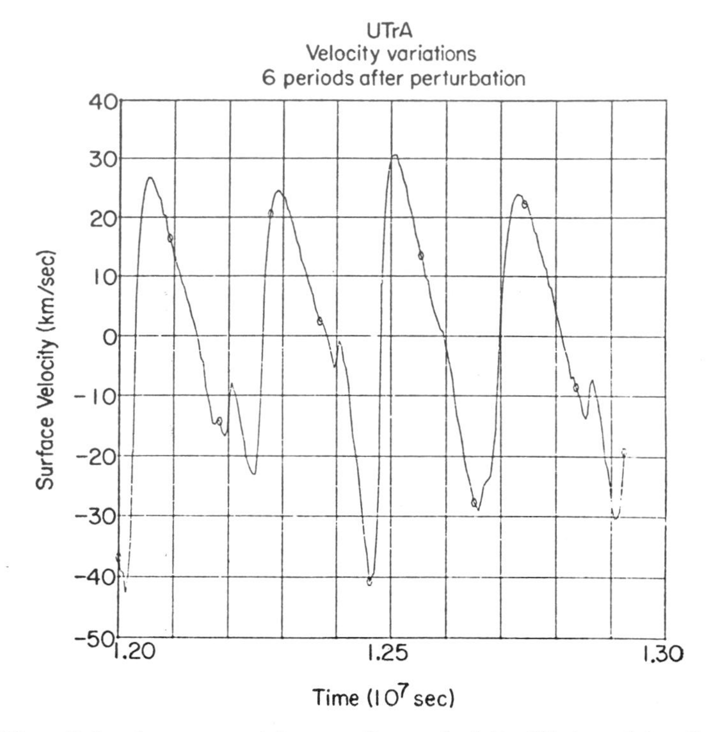

Fig. 11. Velocity curve history for a 1.6 $M_{\odot}$ UTrA model after an artificial first overtone perturbation of 15 percent of the total kinetic energy is put onto a pure fundamental mode.

The situation is confused. We observed mixed-mode Cepheids but we cannot explain their behavior. Observationally and theoretically they must be near the red edge, but are they not low mass? The evidence is strong that they are indeed low mass stars because periods are changed only a few percent by nonlinear theory and period ratios are changed even less. Luminosities are not greatly in error if the radius and effective temperature are fixed. What causes mixed-mode behavior observed in some stars?

Assuming the reality of the low masses, we conjecture that compositions appreciably different from those used heretofore may affect the results of the nonlinear calculations. Composition changes do not greatly affect periods or period ratios, and therefore, masses and radii. The internal composition may affect, however, the detailed nonlinear effects and give

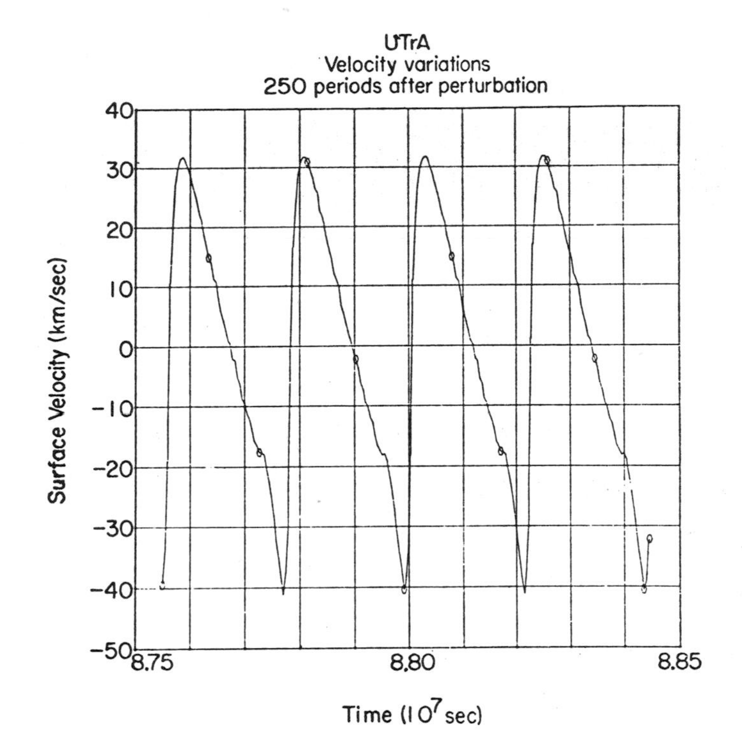

Fig. 12. Velocity curve history for a 1.6 $M_\odot$ UTrA model 250 periods after an artificial perturbation. All traces of the first overtone mode have disappeared.

multiperiodic behavior.

Finally we only mention convection. The energy flow by convection is significant near the red edge of the instability strip, but has been neglected in all studies to date. This, as always, needs immediate attention.

Our conclusion is that we do not yet understand all the parameters of multiperiodic Cepheids, but their masses and radii given by Petersen and more recently by King, Hansen, Ross, and Cox do not seem to be greatly in error. Prediction of observed multiperiodic pulsations has not yet been achieved for Cepheids.

References

Cogan, B. C. 1970, Ap. J., 162, 139.

Cox, J. P. 1974, Repts. Prog. Phys., 37, 563.

Cox, J. P., King, D. S., and Stellingwerf, R. F. 1972, Ap. J., 171, 93.

King, D. S., Cox, A. N., Eilers, D. D., and Cox, J. P. 1975, Bull. A.A.S., 7, 251.

King, D. S., Hansen, C. J., Ross, R. R., and Cox, J. P. 1975, Ap. J., 195, 467.

Ledoux, P. and Walraven, Th. 1958, Handb. d. Phys., vol. 51, ed. S. Flügge (N. Y.: Springer-Verlag).

Petersen, J. O. 1973, Astr. and Ap., 27, 89.

Stellingwerf, R. F. 1974, Ap. J., 192, 139.

Stellingwerf, R. F. 1975, Ap. J., 195, 441.

Stellingwerf, R. F. 1975, Ap. J. (in press).

Discussion to the paper of COX and COX

VAN HORN: I would like to understand more about your comments on convection. First, do I understand correctly that convection has not been included in any of the nonlinear hydrodynamic calculations? Second, I thought Stellingwerf's method allowed him to include convection. Has he not done this?

COX: Stellingwerf has not so far included any allowance for convection, either in establishing the model or in the variation of the flux of energy from the central regions. Actually, Baker and Kippenhahn in 1963 and Iben in 1973 considered convection in establishing the model and considered that the convective flux was not modulated by the gamma and kappa effects which operate on radiation flow and cause the instability. No time dependent convection has ever been used in calculations of model Cepheids.

FITCH: If you have a pulsating star with both P_1 and P_2 excited, but not P_0, would you expect it to be near the red side of the strip?

COX: At present, theory cannot predict where such a star might be found.

Multiple Periodic Variable Stars
IAU Colloquium No.29, Budapest, 1975

MULTIPLE PERIODIC RR LYRAE STARS OBSERVATIONAL REVIEW

B. Szeidl
Konkoly Observatory
Hungarian Academy of Sciences

Introduction

In this review paper I concentrate exclusively on the multiple periodic RR Lyrae stars and I do not wish to discuss the general problems of cluster variables. There are some excellent review papers on them (e.g. Preston, 1964).

The majority of the RR Lyrae stars repeat their light curves from cycle to cycle with astonishing regularity. It was, however, discovered already in the early days of variable star research that some of them showed conspicuous changes in the height of maxima of their light curve and simultaneously the time of maxima could not be fitted in by a linear formula.

Almost 70 years ago, Blazhko (1907) was the first who observed the possible periodic variations in the time of maxima of a short periodic cepheid variable. In the case of RW Dra (87.1906) he found that no constant period could satisfy the time of maxima and one had to postulate periodic changes in the fundamental period with a secondary period of 41.6 days. Later Blazhko carried out further studies on RR Lyrae type stars (Blazhko 1925, 1926) and found that XZ Cygni changed its light curve from cycle to cycle with a secondary period of 57.39 days.

One of the first remarks that the maximum light of some of the RR Lyrae stars might not be constant at different epochs comes from Sperra (1910). Curiously enough he stated that SU and SW Draconis were showing the effect, whereas modern photoelectric observations do not reveal any light curve variation concerning these two stars.

The striking changes in the maxima of the light curve of RR Lyrae which turned out to be periodic was first stated by

Shapley (1916) some 60 years ago. In his fundamental work Shapley wrote: "There can be no doubt that a real irregularity is present. An attempt was made to find a uniform period for the variations that would satisfy all the observations. This failed in part, perhaps because of insufficient data, but it seems that for the whole series the oscillation is roughly periodic with a varying amplitude". He obtained a secondary period of 40 days and an amplitude of 37 minutes for the time oscillation of the median magnitude of the ascending branch of RR Lyrae. These results were perfectly confirmed by Hertzsprung (1922).

Since Blazhko was the very first who demonstrated that the changes in the short periodic light curve could be satisfied by a longer period we usually refer to these periodic variations of the light curve as "Blazhko-effect".

Although many astronomers assailed the problems of multiple periodic RR Lyrae stars during the past 50 years we are still far from solving them. Some excellent papers (Detre 1956, 1962, Klepikova 1956, 1957, 1958, etc.) summarizing the observational results on multiple periodic RR Lyrae stars appeared during the past 20 years. I do not wish to reproduce them, therefore I would like rather to speak about some problems till now not yet discussed and some new observational results.

The first question I deal with is the frequency of multiple periodic RR Lyrae stars.

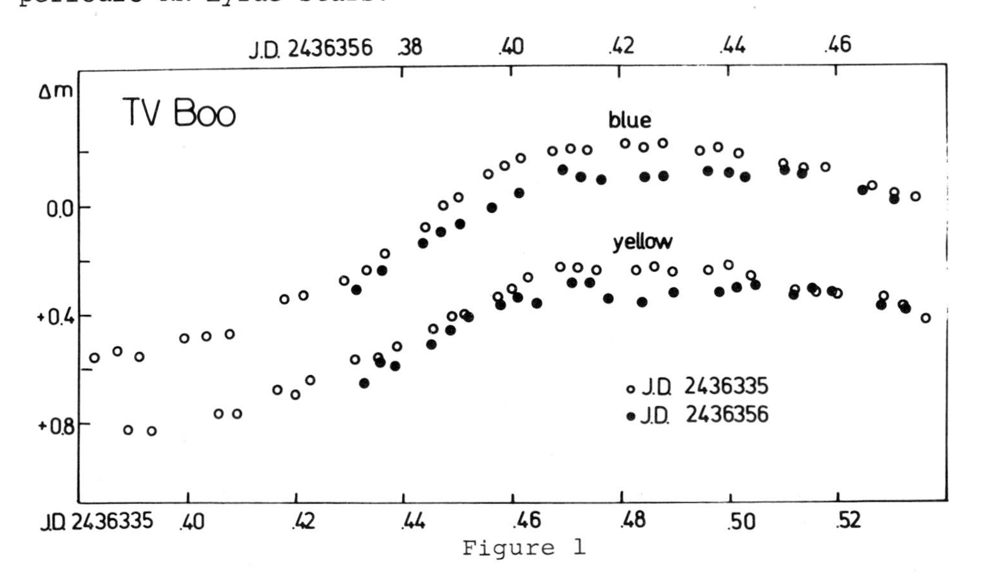

Figure 1

The Frequency of RR Lyrae Stars with Blazhko-effect

Nearly 6000 RR Lyrae stars are listed in the Third Edition of the GCVS and its first and second supplements. We find, however, references to Blazhko-effect only in about 120 cases. It would mean that about 2% of the RR Lyrae stars would have Blazhko-effect. But it is obvious that a statistical investigation of this kind cannot be based on the very inhomogeneous data of the GCVS. In many cases, especially for RR Lyrae stars fainter than 13 magnitudes, the light curve variation remains unnoticed, while on the contrary it may also happen that the large scatter of the observations on the light curve due to observational errors in reality is inadvertently attributed to Blazhko-effect.

The discovery of any variations in the light curve of an RRc type star encounters special difficulties. The changes in the height of maximum are usually small and so accurate photoelectric observations can only settle the question whether an RRc-star has Blazhko-effect or not. Taking into account the observational difficulties mentioned as regards RRc type variables, at present we can only hope that a reasonable estimate might be only given for the frequency of RRab stars with Blazhko-effect.

All the RRab stars with known secondary period are listed in Table 1. The stars are arranged according to the length of the fundamental period, P_O. P_B denotes the length of the secondary period often referred to as Blazhko-period.

In RR Gem the Blazhko-effect ceased to exist around 1940. The same happened in SW And about 1956. Photoelectric observations obtained at the Konkoly Observatory since that time have only shown variations of the hump on the ascending branch of the light curve of SW And. These changes seem to be periodic.

I left out from the list some frequently mentioned RR Lyrae stars with alleged Blazhko-effect. These are XX And, AT And and AN Ser. Our photoelectric observations made during the past 10-15 years have not revealed any light curve variations of these stars.

Now we can turn again to the problem of the frequency of the RRab stars with Blazhko-effect. The paper by Fitch et al. (1966) presents a fairly homogeneous and well-observed sample of RR Lyrae stars. Fifteen of the 90 RRab variables of Fitch et al. are also included in my list (Table 1). They definitely show the Blazhko-effect or used to show it (SW And and RR Gem).

So thus we can state with certainty that about 15-20% of the field RRab variables show light curve variations.

Table 1

RRab Lyrae Stars with Blazhko-effect

Star	P_o	P_B	ΔS	References
RS Boo	$0^d.377$	537^d	2	Oosterhoff BAN 10.101
RR Gem	.397	37	3	Budapest, unpublished
MW Lyr	.398	33.3		Mandel,Per.Zvezdy 17.335
SW And	.442	36.8	0	Balázs,Detre Bp.Mitt.36
RW Dra	.443	41.7	3	Balázs,Detre Bp.Mitt.27
RV Cap	.448	225.5	6	Tsessevich,Astr.Circ.757
TU Com	.461	75		Ureche,Babes-BolyaiStud. Jasc.1.73
TZ Cyg	.466	57.3	6	Muller, BAN 12.11
RV UMa	.468	90.1	8	Preston,Spinrad,Ap.J.147. 1025
AR Her	.470	31.6	6	Almár, Bp.Mitt.51
XZ Dra	.476	78	3	Batyrev,Per.Zvezdy 10.292
X Ret	.492	45:	3	Hoffmeister,Ver.Sonn.5.3.1
V674 Cen	.494	29.5:		Hoffmeister,Ver.Sonn.5.3.1
RZ Lyr	.511	116.7	9	Romanov, IBVS 205
V434 Her	.514	26.1		Rozovsky,Per.Zvezdy 15.211
SW Psc	.521	34.5		Ureche, IBVS 532
Y LMi	.524	33.4		Martynov, Eng.Bull.18
SZ Hya	.537	25.8	6	Kanyó, IBVS 490
UV Oct	.543	80:	9	Hoffmeister, Ver.Sonn.5.3.1
RW Cnc	.547	29.9		Balázs,Detre,Bp.Mitt.23
TT Cnc	.563	89	7	Szeidl, IBVS 278
RR Lyr	0.567	40.8	6	Preston et al.Ap.J.Supp.12.9

Table 1 (cont.)

Star	P_o	P_B	ΔS	References
AR Ser	$0^d.575$	105^d	8	Szeidl, IBVS 220
DL Her	.572	33.6	6	Szeidl, IBVS 36
V365 Her	.613	40.6		Tsessevich, Astr.Zh.38.293
Z CVn	0.654	22.7	8	Kanyó, IBVS 146

Table 2

RRc Lyrae Stars with Blazhko-effect

Star	P_o	P_B	ΔS	References
TV Boo	$0^d.313$	$33^d.5$	8	Detre, Astr.Abh.Leipzig, 1965
BV Aqr	.364	11.6		Tsessevich, Vistas in Astr. vol.13
RU Psc	0.390	28.8	7	Tremko, Bp.Mitt.55

From Table 1 we can also see that the RR Lyrae stars with Blazhko-effect generally have rather low metal abundance. A rough estimate results in the following frequency of multiple periodic field RR Lyrae stars depending on metal abundance:

10% if ΔS=0-2; 20% if ΔS=3-5; 30% if ΔS=6-10

By studying the period changes of RR Lyrae stars Tsessevich (1972) came to the conclusion that "stars with shortage of calcium are characterized by the instability of their pulsations". His result may be connected with the previous statement.

In connection with the ceasing of the Blazhko-effect in SW And and RR Gem (and perhaps in XX And, AT And and AN Ser?) some interesting questions arise. Whether the starting of Blazhko-effect in an RR Lyrae star may happen more than once or all RR Lyrae stars inevitably become multiple periodic at least once during their life when observed with sufficient precision or for long enough intervals of time? Balázs-Detre and Detre (1962) have commented that probably there is no sharp distinction between single and multiple periodic variables. These questions cannot be answered at present. I must, however, emphasize that we have not yet observed commencement of the Blazhko-effect in an RR Lyrae star. Special attention should be paid to the RR Lyrae stars in globular clusters if we investigate the frequency of multiple periodic RR Lyrae stars. One of the best obeserved cluster is Messier 3, (Szeidl 1965, 1972). From the 105 well-observed RRab stars 36 show light curve variations, i.e. 35% of the variables. In Table 3 are given the frequency distribution according to period. The highest frequency occurs around P=0.56.

Table 3

P ±0.015	all	with Blazhko-effect	%
0.47	8	3	37.5
0.50	24	8	33.3
0.53	26	10	38.5
0.56	15	8	53.5
0.59	16	4	25.0
0.62	7	2	28.6
0.65	6	1	16.7
0.68-	3	0	0

The investigation of cluster variables in M15 has led to similar results. In this cluster the frequency of RR Lyrae stars with Blazhko-effect is about 25% /Barlai, 1975/.

The higher frequency of RRab stars with Blazhko-effect in clusters is well understandable and is in good agreement with the fact that this kind of RRab stars in the field are more frequent among the metal poor stars.

Characteristics of Multiple Periodic RR Lyrae Stars

It was an open question for a long time whether RRc-type stars could have Blazhko-effect or not. Nevertheless it has been known that the light curve of some of them is unstable. Especially photoelectric measurements showed this phenomenon clearly. As an example I mention the RRc type variable T Sextantis for which Tifft and Smith (1958) found definite light curve variation particularly at maximum and around the hump.

Now we definitely know three RRc stars with periodic changes of their light curves (see Table 2). One of them is TV Bootis. It was invesitgated by Detre (1965) who found its secondary period to be 33.5 days. Figure 1. shows the two extreme forms of maxima in yellow and blue light. The hump is very pronounced if the maximum is fainter. The amplitude of the variation in the brightness of the maximum is 0.12 magn. in blue and 0.09 magn. in yellow light while the oscillation of the ascending branch is only $0^{d}.01$. These data indicate that the changes are small and can only detected by photoelectric observations.

Ten years ago or so Tremko (1964) analyzed the long series of observations of RU Psc. He was able to construct the O-C diagram of the star which shows that the fundamental period of RU Psc has undergone very rapid and complicated changes. The variation in its light curve is also small and is similar to that of TV Boo. Tremko derived a secondary period of 28.8 days for RU Psc.

The third known RRc star which shows Blazhko-effect is BV Aqr. Tsessevich obtained a secondary period of 11.6 days for this star. This value is the shortest secondary period ever observed for an RR Lyrae star.

Generally the RRab stars with Blazhko-effect show large changes in their light amplitudes. At present we know 27 RRab stars with secondary periods (Table 1). About half of them are thoroughly investigated (see e.g. Almár 1961, Klepikova, Balázs and Detre 1943, 1950, 1952, 1957) I do not wish here to speak of all of them in detail, I should like only to stress some points I think to be important.

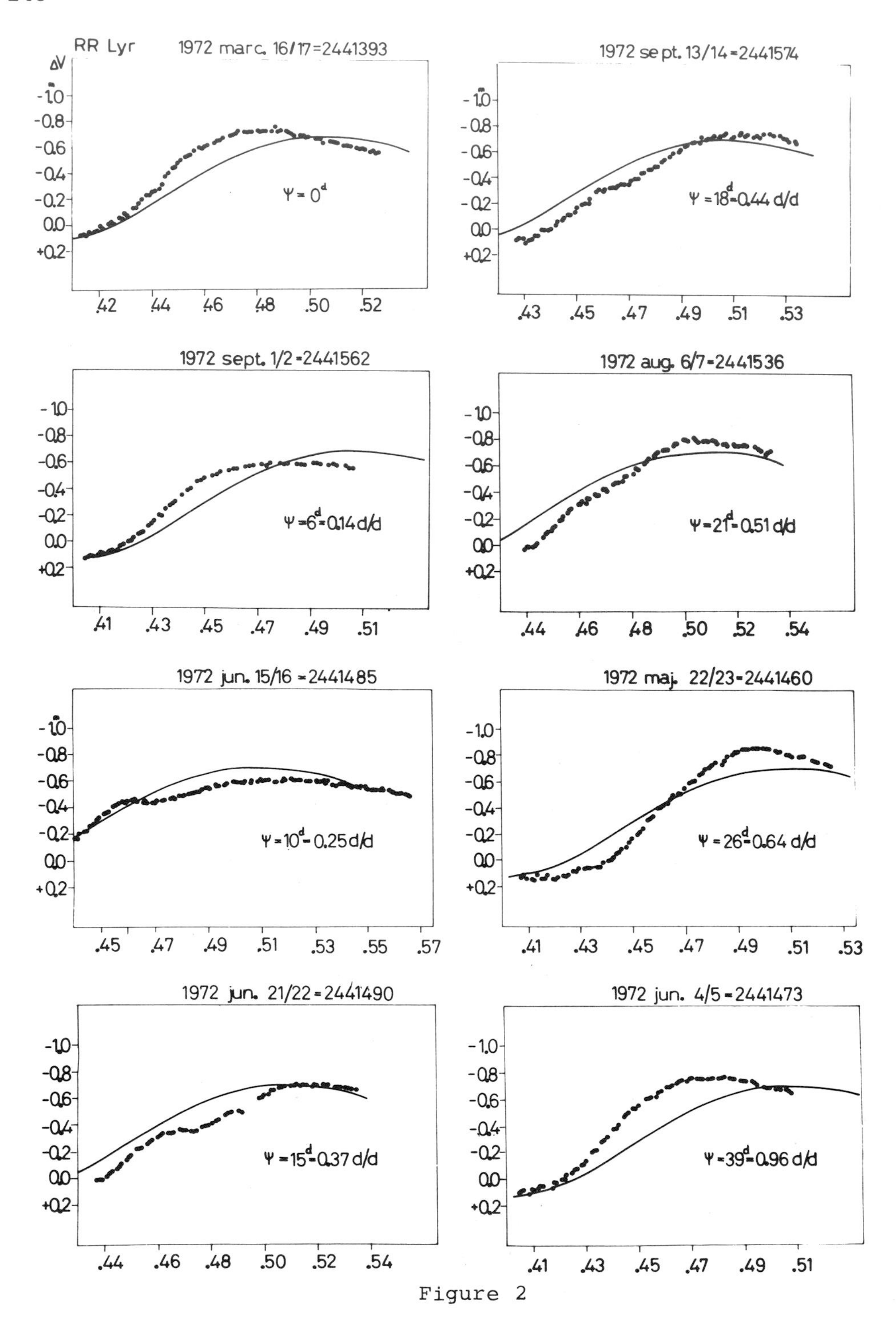
RR Lyr
ΔV
1972 marc. 16/17=2441393
Ψ = 0^d
1972 sept. 13/14=2441574
Ψ = 18^d=0.44 d/d
1972 sept. 1/2=2441562
Ψ = 6^d=0.14 d/d
1972 aug. 6/7=2441536
Ψ = 21^d=0.51 d/d
1972 jun. 15/16=2441485
Ψ = 10^d=0.25 d/d
1972 maj. 22/23=2441460
Ψ = 26^d=0.64 d/d
1972 jun. 21/22=2441490
Ψ = 15^d=0.37 d/d
1972 jun. 4/5=2441473
Ψ = 39^d=0.96 d/d

Figure 2

It is remarkable that a part of the ascending branch of the light curve appears to be substantially constant in time and phase at different phases of the secondary period. If the secondary period is relatively short, this constant part occurs on the lower portion of the ascending branch and the phase variation is greatest at maximum (e.g. RW Cnc). If the secondary period is relatively long, the time oscillation of the lower portion of the rising branch has a fairly large amplitude (e.g. XZ Dra) (Detre, 1962). All this is, however, not a general rule. RR Lyrae itself is the brightest known of the cluster type variables having Blazhko-effect and therefore specially fits for detailed investigation. From Figure 2 we can get a rough idea how the light curve, the maximum, the hump and steepness of the ascending branch change and develope during the secondary cycle. In the Figure, 8 diagrams are given showing our separate photoelectric observations of the ascending branch of the light curve and the adjacent minimum and maximum on different nights at different ψ phases of the secondary period. For comparison the mean light curve has been drawn on each diagram. The light curves show the usual particulars of stars of this kind.

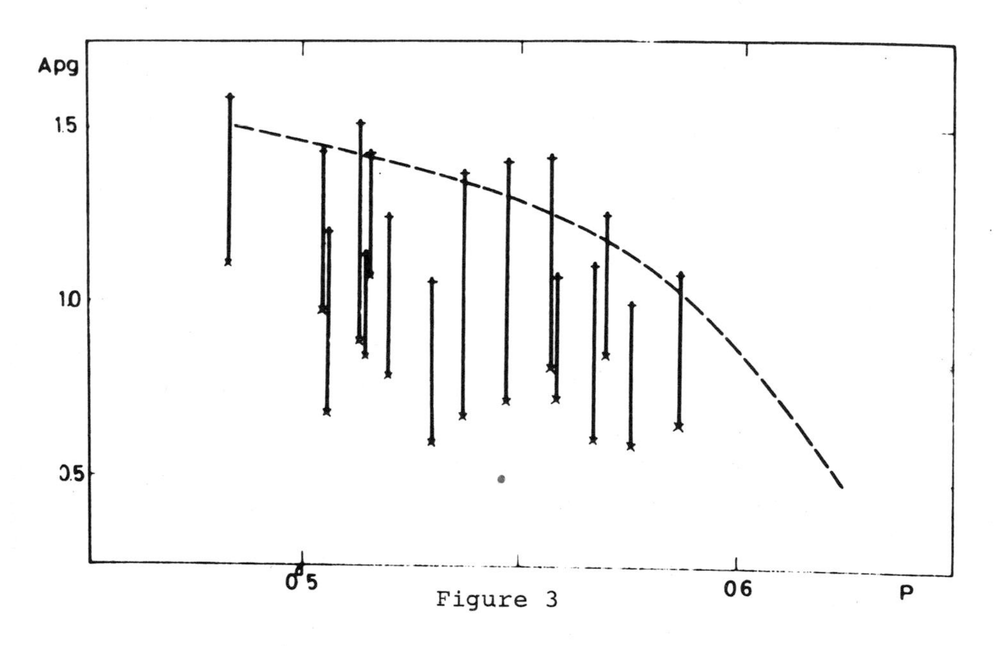

Figure 3

In the same star, different cycles of the Blazhko-effect differ in length and amplitude as was first shown by Walraven (1948) for RR Lyr itself. In some cases (e.g. RW Cnc) these differences between successive cycles are especially striking. Attempts were made at representing these variations by longer periods of several months but they were not succesful. The reality of the reported additional periods of RR Lyrae and other variables were challanged (Detre, Balázs-Detre 1962). I am of the opinion that these variations are not periodic.

Sometimes the Blazhko-effect undergoes considerable changes. The most interesting common feature of all the RRab stars with Blazhko-effect is, however, that during these considerable changes the brightness of the highest maximum remains substantially constant while the lowest maximum varies (e.g. Almár, 1961). This result is in accordance whit that obtained for RR Lyrae stars in Messier 3. Figure 3 shows that the greatest amplitudes of variables with Blazhko-effect fit the period-amplitude relation valid for RR Lyrae stars with stable light curves. In some cases the greatest amplitude is below the expected value, probably because only smaller amplitudes have been observed at Budapest.

Figures 4 a,b,c show the variations of brightness and phase of maximum light of RR Lyrae in the years 1972, 1973 and 1974, respectively. Along the loops the corresponding phases of the secondary period are also given. The form of the loops changes insignificantly in the course of time.

In Figures 5 a-e the brightness variation and phase oscillation of the maxima of the light curves are separately plotted against the phase Ψ of the secondary period for some RR Lyrae stars with Blazhko-effect. It is immediately conspicuous that the phase shift $\Delta\Psi$ between the highest and the most positive shifted maximum is characteristic of multiple periodic RR Lyrae stars. Table 4 gives the $\Delta\Psi$ values of some RR Lyrae stars with Blazhko-effect. I am unable to find any connection between $\Delta\Psi$ and the fundamental or secondary period.

Figure 4

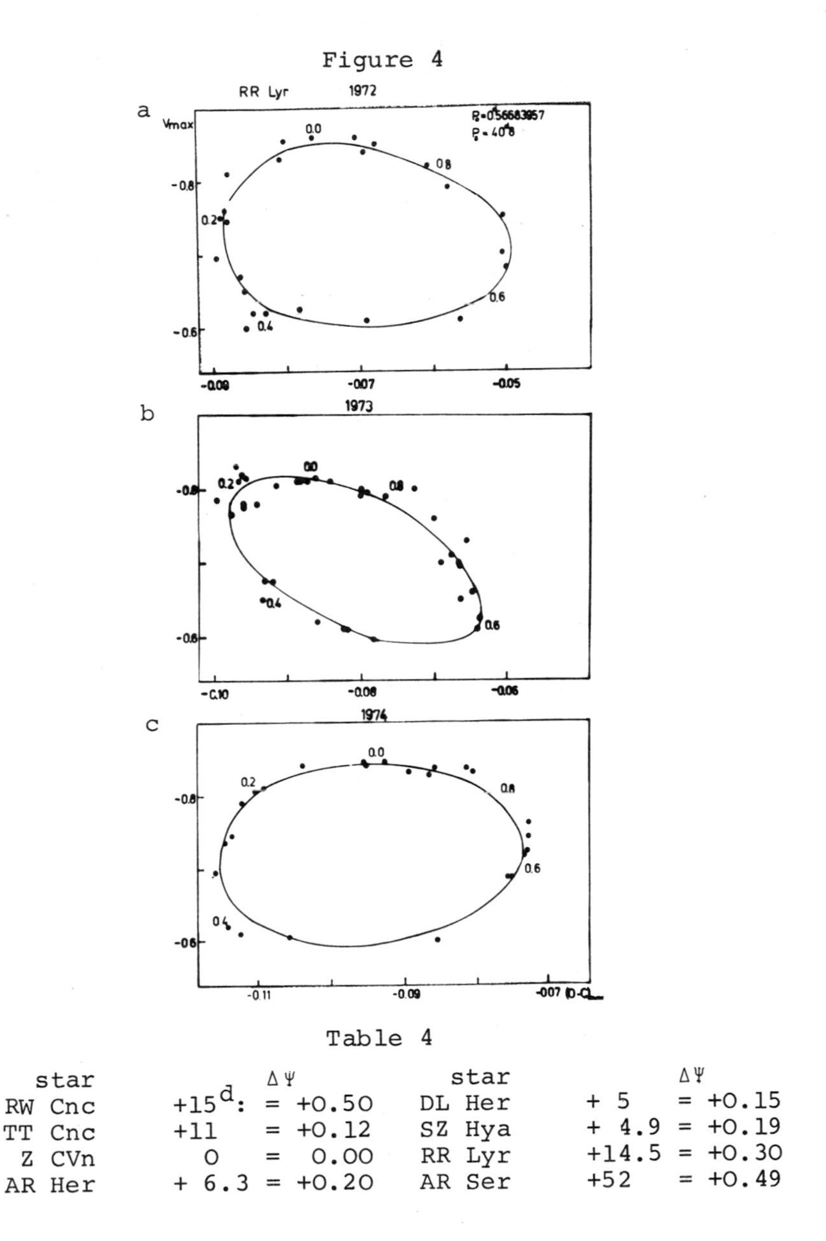

Table 4

star		$\Delta\Psi$	star		$\Delta\Psi$
RW Cnc	$+15^{d}$:	= +0.50	DL Her	+ 5	= +0.15
TT Cnc	+11	= +0.12	SZ Hya	+ 4.9	= +0.19
Z CVn	0	= 0.00	RR Lyr	+14.5	= +0.30
AR Her	+ 6.3	= +0.20	AR Ser	+52	= +0.49

The most interesting result on the Blazhko-effect of RR Lyrae was obtained in the last few years. At the 1968 Variable Star Colloquium Detre (1969) presented a figure based on our photoelectric observations representing the phase variation of the median brightness on the rising branch and the variation in the height of the visual maximum of RR Lyrae in the course of the 41-day secondary cycle from year to year between 1962 and 1968. In 1963 and 1967

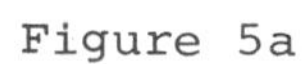

Figure 5a

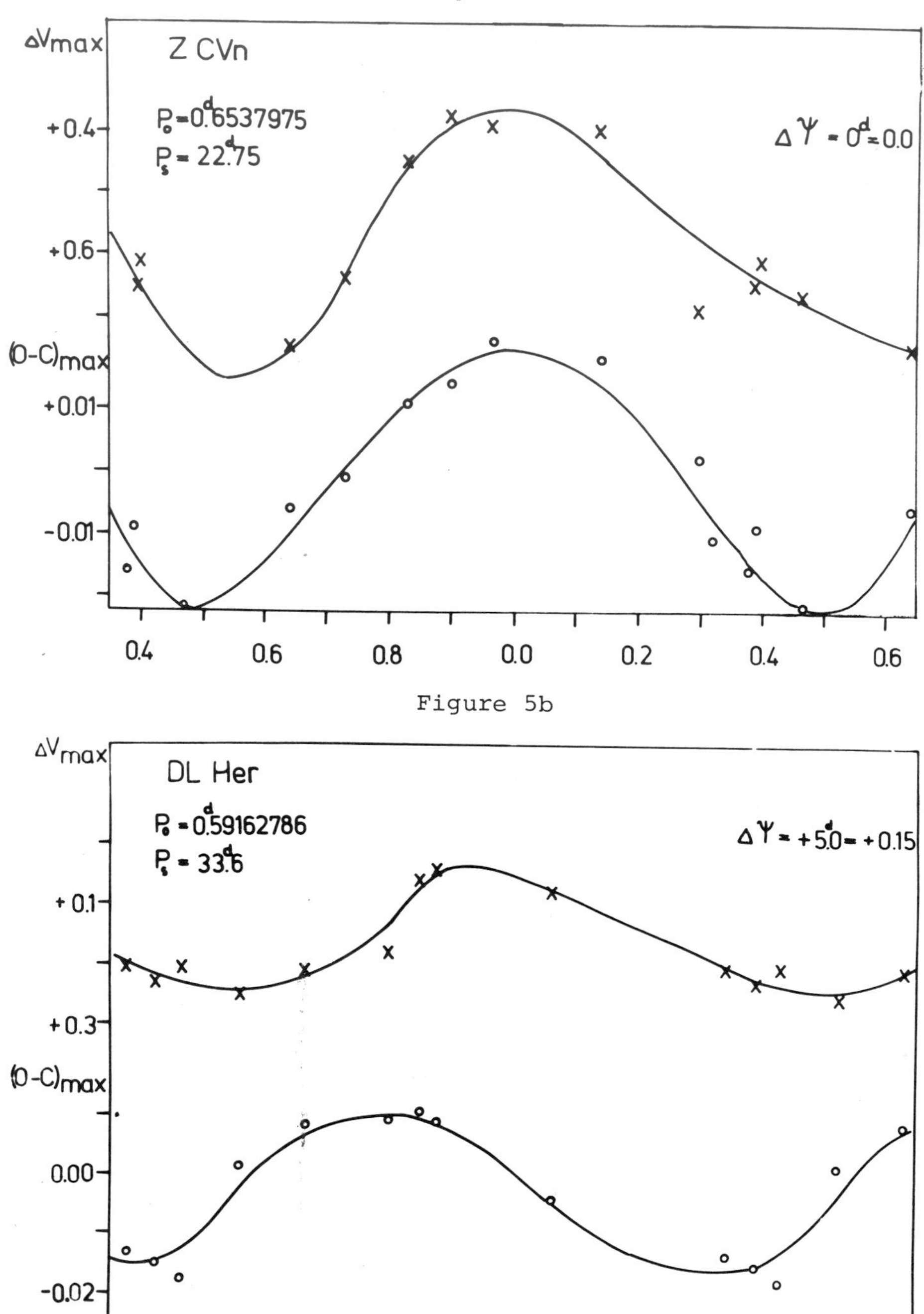

Figure 5b

Figure 5c

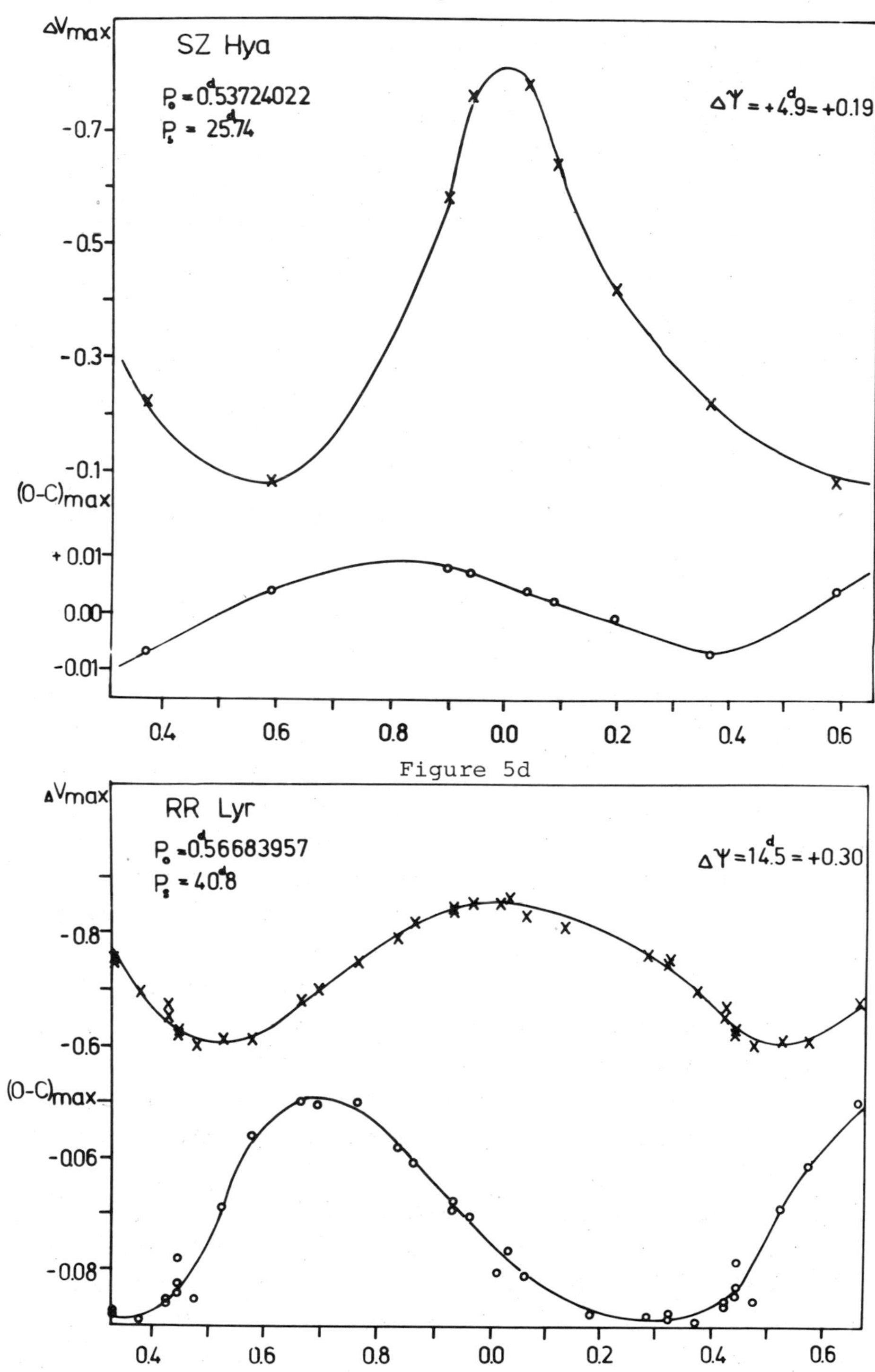
ΔVmax
SZ Hya
P0 = 0.53724022
Ps = 25.74
ΔΨ = +4.9 = +0.19
-0.7
-0.5
-0.3
-0.1
(O-C)max
+0.01
0.00
-0.01
0.4
0.6
0.8
0.0
0.2
0.4
0.6
Figure 5d
ΔVmax
RR Lyr
P0 = 0.56683957
Ps = 40.8
ΔΨ = 14.5 = +0.30
-0.8
-0.6
(O-C)max
-0.06
-0.08
0.4
0.6
0.8
0.0
0.2
0.4
0.6

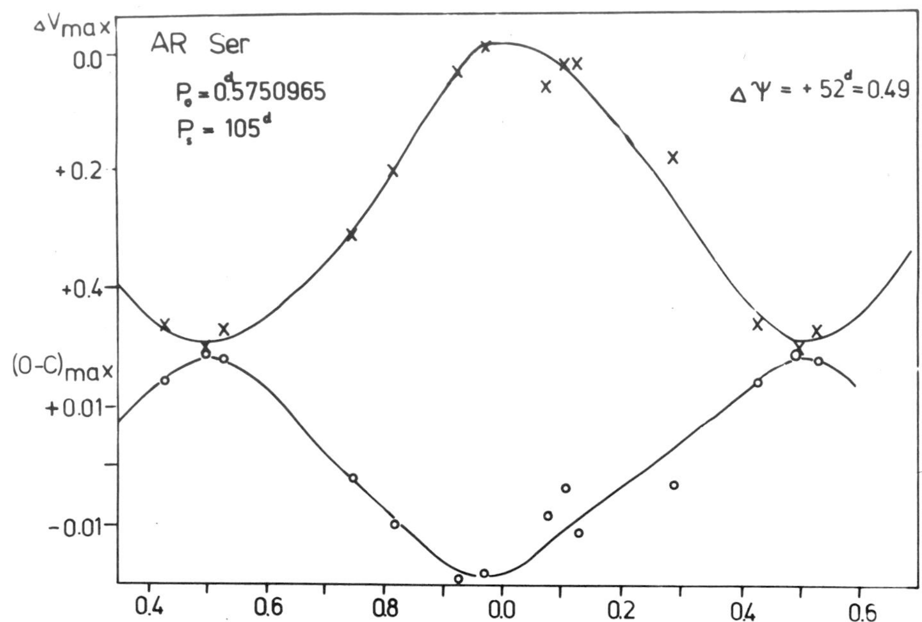

the variations connected with the 41-day cycle were very small, while they were strong in other years. In 1971 the amplitudes were very small again. In this way a 4-year cycle became apparent. To our great satisfaction, this year we were also able to observe the transition from an old 4-year cycle to a new one. Thus the existence of the 4-year period is definitely established. Discussing old visual and photographic observations we were able to follow this long period cycle back to 1935 (Detre and Szeidl, 1973). In Figure 6 the phase variation of the median brightness on the ascending branch and the variation in the height of the visual maximum of RR Lyrae are given during the last 4-year cycle.

At the end of an old 4-year cycle the amplitude of the maximum variations is smaller than 0.1 magn., and then very rapidly becomes as large as 0.2-0.3 magn.

Most interesting is the phase-shift in the 41-day period following the transition from an old 4-year cycle to a new one. The O′-C′ value of the most positive-shifted ascending branches was +19 days throughout 1967-71, +29 days during 1971-75, while for the new cycle it is +19 days, i.e. the beginning of a new cycle is accompanied by a phase shift of 10 days, about a quarter of the 41-day period. After each discontinuity the O′-C′

Figure 6

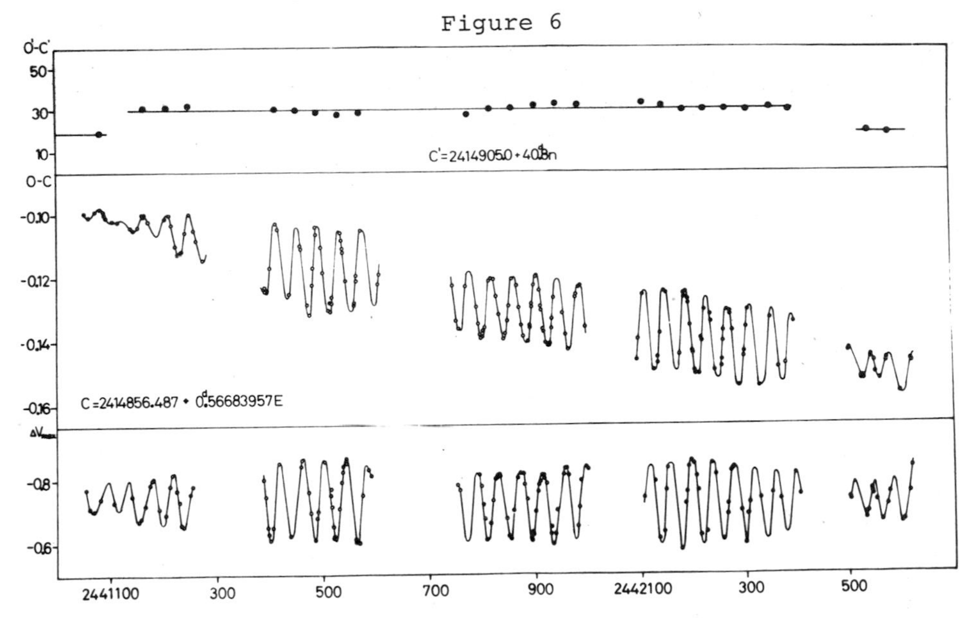

of the secondary period remains constant during one and the same 4-year cycle, during the past 40 years only a very strong cycle seems to be an exception.

Long period variations in the Blazhko-effect are known for other RR Lyrae type variables, as well.

These are :	XZ Cyg	$P_1 \approx 9$ years	(Klepikova, 1958)
	RV UMa	7 years	(Kanyó, 1975)
	RW Dra	10 years	(Budapest,unpublished)

These long periods may remind us of the solar cycle. The magnetic field intensity of one cluster type variable was only measured. Babcock (1958) obtained 20 measures for RR Lyrae itself. If we arrange these observations according to the phases of the main (P_o) and secondary period (P_B) we do not find any correlation with the main period but a separation of positive and negative values exists in the course of the 41-day secondary period as was shown by Detre (1962). Brightest maxima appear to coincide with largest negative,lowest maxima with the largest positive values of the magnetic field intensity. Preston's magnetic observations of RR Lyr (Preston, 1967) in 1963 coincide with the minimum of a 4-year cycle, while those in 1964 with the beginning of a weak 4-year cycle. That may be the explanation of why he could not find a

Figure 7

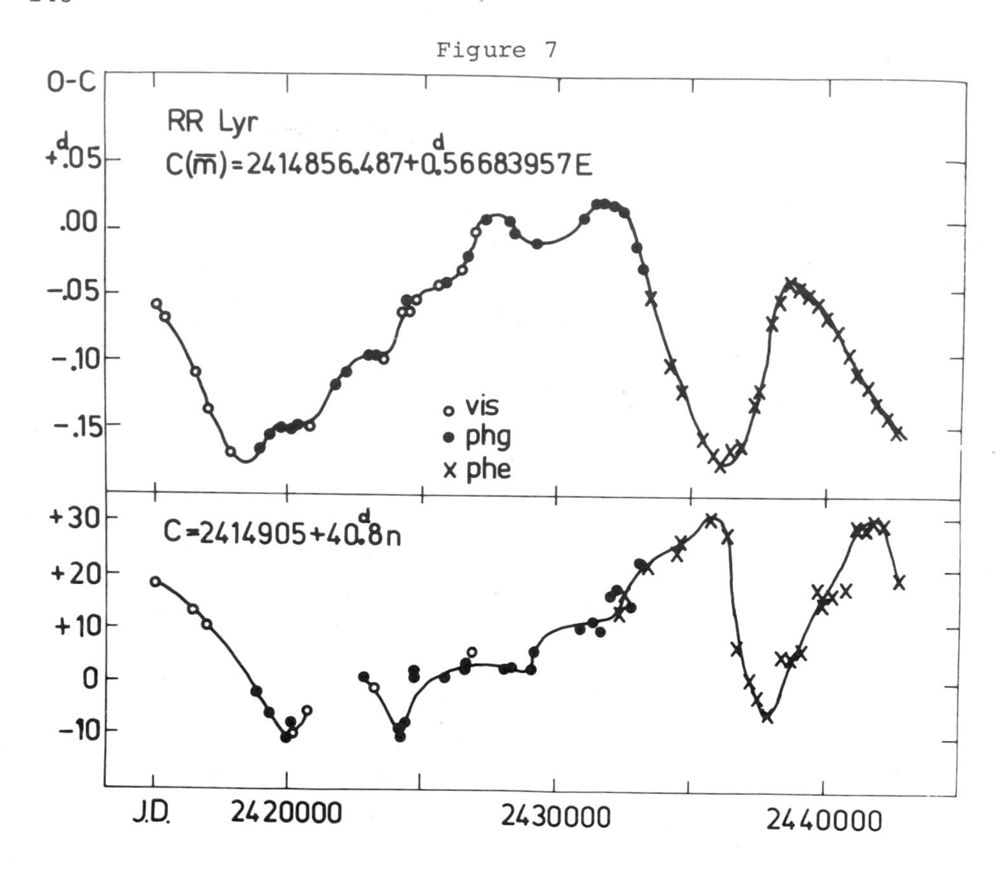

measurable field. On the bases of all these we suggest that the Blazhko-effect of RR Lyrae stars may be connected with their magnetic field. If true,it may be of fundamental importance in understanding the nature of Blazhko-effect.

Briefly I should like to mention the problems of O-C diagrams of multiple periodic RR Lyrae stars. The O-C diagrams of the main period of this kind of variables are usually very complicated (Szeidl, 1965). In Figure 7 the O-C diagram of RR Lyrae is given. The O′-C′ diagram for its secondary period is also drawn in. The two diagrams seem to be the mirror image of each other. This is characteristic of RR Lyrae stars with Blazhko-effect (Ustinov, 1951; Balázs and Detre, 1962), but it is not a general rule. As can be seen from Figure 7 the main and secondary period of RR Lyrae have not always changed in opposite direction.

References

Almár, I., 1961, Mitteilungen Sternw.Ung.Akad.Wiss.Nr.51
Babcock, H.W., 1958, Astroph.Journ.Suppl.3. 141
Balázs, J.and Detre, L.,1943,Mitt.Sternw.Ung.Akad.Wiss.Nr.18
1950, ibid, Nr.23
1952, ibid, Nr.27
1957, ibid, Nr.34
1962, Kleine Veröff.Remeis-Sternwarte Bamberg,Nr.34.p.90
Blazhko,S., 1907, Astron.Nachr. 175. 325
1925, Ann. de l'Obs.Astr.de Moscow, Sér.2 vol.VIII. livr.No.1
1926, ibid, No.2
Detre, L., 1956, Vistas in Astronomy, Vol.2, p.1156
1962, Trans.I.A.U. vol.XIB-Proceedings,p.293
1965, Astron.Abh.Leipzig, p.621
1969, Non-Periodic Phenomena in Variable Stars, Ed. Detre, p.3
Detre, L. and Szeidl, B.,1973, Variable Stars in Globular Clusters and in Related Systems, Ed.Fernie,p.31
Fitch, W.S., Wisniewski, W.Z.and Johnson, H.L., 1966, Comm. Lunar and Planetary Lab.No.71, Vol.5,Part 2
Hertzsprung, E., 1922, BAN 1, 139
Kanyó, S., 1975, Mitt.Ung.Akad.Wiss. Nr. 69
Klepikova, L.A., 1956 Peremennye Zvezdy 11, 1.
1957 ibid. 11. 137
1958 ibid. 12. 164
Preston, G.,1964, Annual Review of Astronomy and Astrophysics, vol.2, pp.23-48
1967, The Magnetic and Related Stars, Ed.Cameron, Mono Book Coop. Baltimore, p.26
Shapley, H., 1916, Astroph. Journ. 43, 217
Sperra, W.E., 1910, Astron. Nachr. 184, 241
Szeidl, B., 1965, Mitt.Sterw.Ung.Akad.Wiss. Nr.58
1974, ibid. Nr.63
Tifft, W.G. and Smith, H.J., 1958, Astroph.Journ.127.591
Tremko, J.,1964,Mitt.Sternw.Ung.Akad.Wiss. Nr. 55

Tsessevich, V., 1972, Vistas in Astronomy Ed. Beer, Pergamon Press, vol. 13, p. 241

Ustinov, B.A., 1951, Astr. Circ. No. 110

Walraven, Th., 1948, BAN, 11. 17

Discussion to the paper of SZEIDL

FROLOV: Does the U-B versus B-V loop always have the same counter-clockwise direction for RRab stars with a Blazhko-effect, or can the direction of circulation sometimes change with the phase of the Blazhko-effect?

SZEIDL: I have not found that the direction changes with phase of the Blazhko-effect.

DZIEMBOWSKI: Tsessevich has suggested some time ago that the radial velocity curves of some RR Lyrae stars with Blazhko-effect may be interpreted as a beat phenomenon of two oscillations with close frequencies. Could you comment on this?

SZEIDL: In my opinion, the radial velocity curves are not accurate enough. I do not think that any modern models can account for the period ratio in the case of RR Lyrae stars with a Blazhko-effect.

DZIEMBOWSKI: But if we admit the possibility of nonradial pulsation, we cannot rule out the above interpretation on theoretical grounds. Are there any direct observational evidences against it?

SZEIDL: I think there are. I mention two. There is always a point on the ascending branch of a multiperiodic variable which does not vary during the secondary period. There is another interesting fact. A new 4-year cycle of RR Lyrae always begins with a phase shift in the 41-day period.

TREMKO: The Blazhko-effect was present in RU Psc both in the period and in the light variations. The variations of the primary period in the past were also found, and thus the long period variations of the Blazhko-effect are to be expected.

WOLFF: I was interested in your comments on the possible reason for the discrepancy in the magnetic field measurements of Babcock and of Preston. If your hypothesis is correct when would you expect the field to be detectable again?

SZEIDL: Preston's magnetic measurements were made in 1963 and in 1964. In 1963, the Blazhko-effect of RR Lyrae was very weak. The 4-year cycle was in minimum. Then a weak new cycle developed. If the hypothesis is correct, I expect the field to be detectable in the next 3 years. I expect a new minimum in the Blazhko-effect around 1979.

Multiple Periodic Variable Stars
IAU Colloquium No.29, Budapest, 1975

THEORY OF MULTIPERIODIC RR LYRAE STARS

R. F. Stellingwerf

Department of Astronomy, Columbia University, New York, N. Y.

1. Introduction

The subject "Multiperiodic RR Lyrae Stars" encompasses two entirely different types of phenomena: 1) long-term modulation of amplitude known as the Blazhko-effect and 2) simultaneous excitation of several radial modes. In the first category the observational material is now quite extensive, but very little theoretical analysis has been undertaken. In the second case the results of pulsation theory are applicable, but only one such object has been found. Much progress in our understanding of RR Lyrae stars has been made in recent years, however, and many results do have a bearing on the present problem. Several excellent reviews of current pulsation theory are available (see Iben 1971c, or Cox 1974a,b). I will therefore restrict my comments to those results bearing directly on choice of mode for RR Lyrae stars.

Many suggestions have been made as to the mechanism of the Blazhko-effect; among them we have:

1) Resonance effects in radial modes (Kluyver, 1936),
2) Resonances involving nonradial modes (Ledoux, 1951),
3) Splitting of radial modes caused by nonadiabatic effects (Ledoux, 1963, p. 421),

4) Tidal effects (Fitch, 1967, 1968),

5) Oblique rotator effects (Balázs, 1959; Preston, 1964), and

6) Stellar magnetic cycle effects (Detre and Szeidl, 1972).

Note that a stellar magnetic cycle does not necessarily imply an oblique rotator since variations during a rotation period could be caused by other non-spherical effects, such as star-spots. The resonance and splitting suggestions for radial modes can now be ruled out since detailed nonlinear nonadiabatic models show no trace of such behavior. Also, Lucy (1975) has shown that a careful analysis predicts no splitting due to low order non-adiabatic effects. Resonance with a nonradial mode is unlikely to occur over the considerable range of observed periods. Detailed theoretical tests of the remaining hypotheses have not been attempted; in view of the complexity introduced, this is likely to remain an observational problem for the near future.

2. Linear Results

Although a linear stability analysis cannot predict full amplitude behavior, this technique provides much valuable information and is subject to less computational uncertainty than the nonlinear approach. In particular, it is likely that all observed modes are unstable in the linear theory. It is clear, however, that all unstable modes are not necessarily present in the final motion.

The first detailed models of this type were computed by Baker (1965). Models with a range of mass, effective temperature and composition at luminosities near $M_{bol} = 0.5$ were computed in this investigation. Convection was included in the static envelope, but convective variations were ignored. Baker found that his periods supported the suggestion by Schwarzschild (1940) that Bailey type a and b stars are fundamental pulsators, while those of type _c_ prefer the first overtone (sometimes loosely referred to as "first harmonic"). Globular cluster variables show a rather abrupt switch of modes as a function of period. Baker's stability analysis indicated, however, that both modes are unstable over a wide range of effective temperature. This is shown in Figure 1 (Baker's Figure 6) where growth rate is plotted versus effective temperature for the fundamental mode--solid line, and the first overtone--dashed line, of a typical sequence. An estimate of the location of the variables in M3 is also shown. No hint of a mode switch is indicated. A continuation of this investigation and a detailed comparison with observations has been given by van Albada and Baker (1971a).

Baker's results may be compared with those of Castor (1971), who computed purely radiative models with mass $M = 0.58\ M_{\odot}$, bolometric magnitude $M_{bol} = 0.76$, helium abundance $Y = 0.3$ and metal abundance $Z = 0.002$. In Figure 2 (Castor's Figure 5) the growth rate η --fractional change in kinetic energy per period--is shown for the first three modes. The results are similar to Baker's except that the radiative instability strip extends to cooler models. Note that the second overtone is always stable.

An extension of these results to a very wide range of parameters has been reported by Iben and collaborators (Iben and Huchra 1970, 1971; Iben 1971a,b; Tuggle and Iben 1972, 1973). One result of interest here is the location of the blue edge of the instability strip in the H-R diagram. As shown in Figure 3 (Figure 19 of Iben 1971c) separate edges are found for the two lowest modes and the relative position of these edges depends on luminosity. At high luminosities (for a given mass) the fundamental mode dominates--as in W Virginis stars; at lower luminosities the first overtone growth rate is somewhat larger than the fundamental growth rate causing the first overtone blue edge to move to the left, and creating a region in

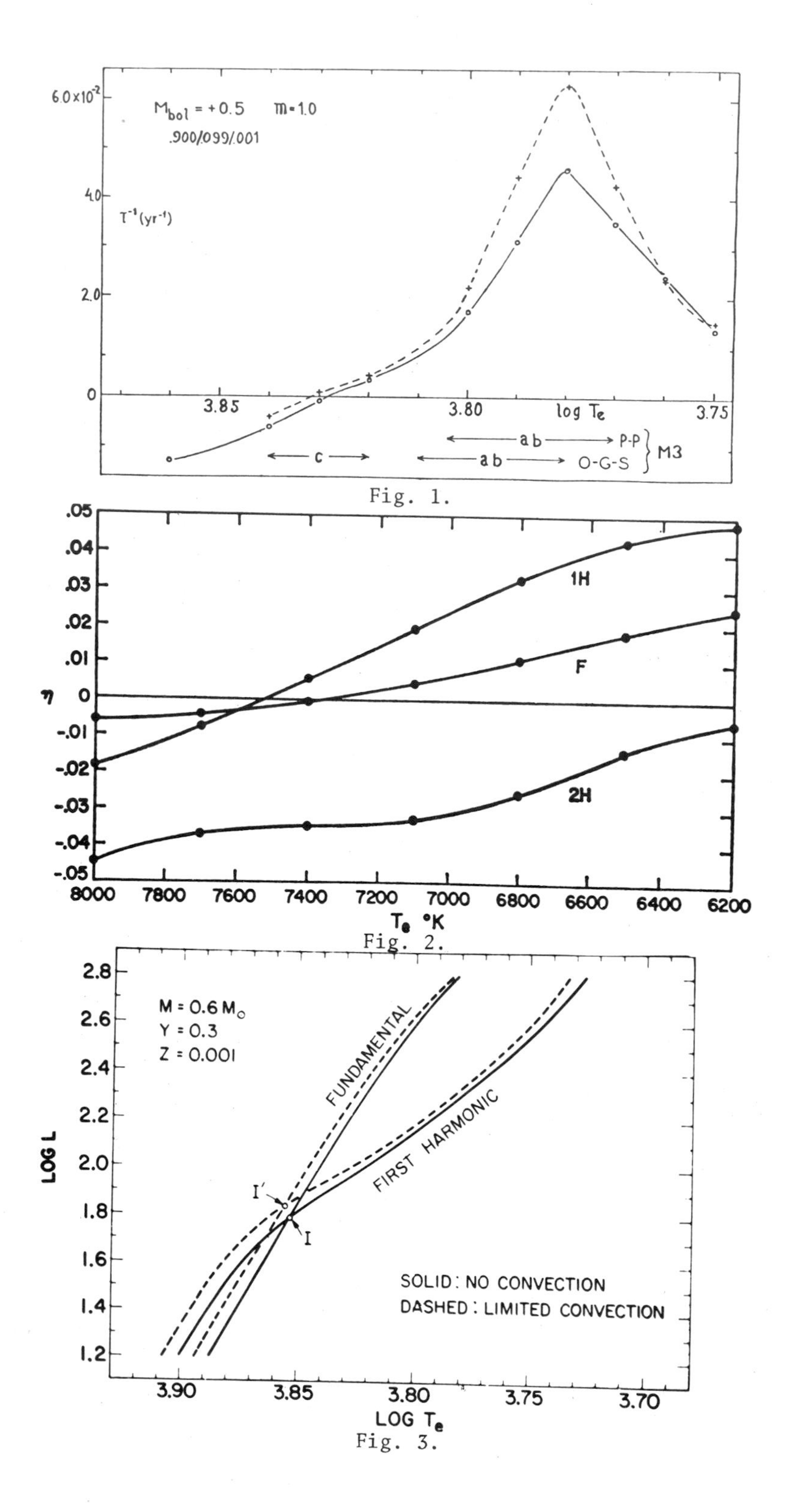

Fig. 1.

Fig. 2.

Fig. 3.

which the first overtone is the only possible mode. At still lower luminosities higher modes can also become unstable (Castor, 1970). If the mass is increased, the point of intersection of the two edges shifts to higher luminosities, while an increase in helium abundance causes a shift to higher temperatures. It has been shown by Cox, Castor and King (1972) that the reduction in excitation of higher modes at high luminosities is caused by the proximity of the outermost node to the driving region. The movement of the node appears to be a function of the parameter M/R, where R is the radius. The first overtone excitation is reduced when, in solar units, $M/R < 0.1$, and the period ratio P_1/P_0 is also reduced (Cox, King and Stellingwerf, 1972). For the second overtone both the growth rate and the period ratio P_2/P_1 (as well as P_2/P_0) decrease rapidly when $M/R < 0.15$. We may therefore define the "dominant linear mode" as that mode with the largest growth rate per unit time at small amplitudes. Figure 4 shows the dominant mode on the H-R diagram for $M = 0.58\ M_\odot$, $Y = 0.3$, $Z = 0.001$ (taken from the results of Stellingwerf, 1975 and unpublished results). Also shown are the regions in which only one mode is excited, and approximate luminosities and temperatures for the variables in M3. Note that for a horizontal branch at $M_{bol} > 1.0$, this figure predicts second overtone pulsation toward the blue. This additional mode does appear in observations of the cluster M68 (van Albada and Baker, 1973).

The relationship of the linear results to the observed modal behavior is discussed by Iben (1971b) and Cox, Castor and King (1972), but in general these predictions have not been born out by the nonlinear calculations. In particular, the dominant linear mode is not invariably the final preferred mode. In the case of stars showing a mixture of modes, however, we may be witnessing the process of approach to the preferred mode (Fitch, 1970) and in this case the dominant linear mode will temporarily be seen.

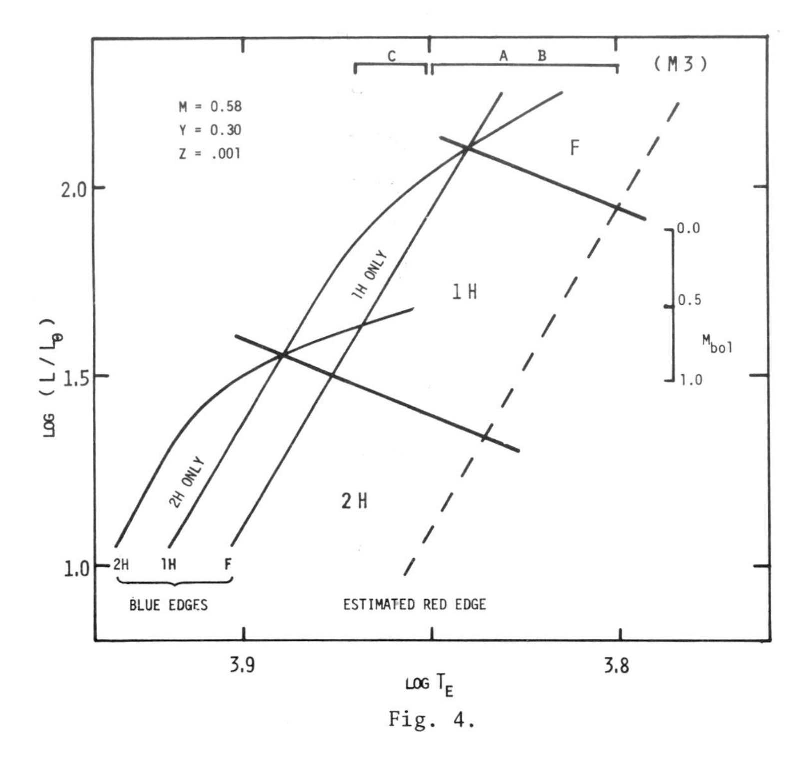

Fig. 4.

The adjustment time as given by the linear models is about 2 months for a 0.6 $M_\odot$ model. If the mass is increased to 3.0 $M_\odot$--appropriate for a star in the first crossing of the instability strip--growth time for the fundamental increases to 5 years only, and is even smaller for higher modes (Stellingwerf, 1975c). This suggests that the observed mixture of modes represents a stable nonlinear configuration rather than a transient coexistance of quasilinear modes.

3. Nonlinear Results

An investigation of modal behavior in nonlinear pulsation models requires an extensive survey over several parameters. The definitive work of this type is due to Christy (1962, 1964, 1966). Christy's approach involves integrating the nonlinear set of equations until the motion appears to be approximately periodic--usually about fifty periods. If the final motion is reasonably stable, Christy argued, it should represent the long-term preferred behavior of the model. Christy was also able to outline the limits of the instability strip by following models at moderate (half maximum) amplitude for only a few periods to determine whether the model showed amplitude growth.

Three types of modal behavior were encountered, 1) stable fundamental pulsation, 2) stable first overtone pulsation, and 3) stable pulsation in either the fundamental or the first overtone ("either-or" behavior). On the H-R diagram the fundamental region occurred at lower effective temperatures than the first overtone region, with the either-or behavior at the fundamental/first overtone boundary. Furthermore, at high luminosity the fundamental behavior is preferred as opposed to first overtone pulsation at low luminosities. The period at the center of the either-or region satisfies the relation

$$P_{tr} = 0.057 \, (L/L_{\odot})^{0.6} \quad . \tag{1}$$

This relation, as projected on the H-R diagram, will be referred to as the "Christy transition line." The width of the either-or region was found to be between 300 K and 600 K in effective temperature (note Christy's seq. 5). This behavior is qualitatively similar to the linear results shown in Figure 4, but the slope of Christy's transition line on this diagram is five times larger than the slope of the R = constant lines dividing the regions of different dominant linear mode.

It should be noted that uncertainties exist with this approach that may not be predictable. The final behavior of a model depends on initialization and total integration time. In fact, the long-term continuation of a calculated mode must always be regarded as problematical.

Spangenberg (1974) has repeated Christy's calculation for $M = 0.6\ M_\odot$, $Y = 0.3$, $Z = 0.001$ using models with improved outer boundary condition and opacity. The uncertainties mentioned above were reduced by extending the integration time to several hundred periods. These results resemble Christy's with the exception of a much wider either-or region, including nearly the entire instability strip at some luminosities. Actual mode transition will occur at the edges of this region, so Christy's transition line--which describes the center of this zone--should be used with caution.

Since the nonlinear models reproduce observed parameters very accurately, it would appear that the problems associated with mode of pulsation are due primarily to uncertainties in the mathematical approach rather than uncertainties in the models themselves. One major difficulty, identification and characterization of the nonlinear limit cycles, can be avoided by using a scheme devised by Baker and von Sengbusch (1969; see also von Sengbusch, 1973; Stellingwerf 1974). This scheme is essentially a generalized Newton-Raphson relaxation and allows the calculation of exact periodic limit cycles. Stability coefficients for perturbations toward other limit cycles are also obtained through application of the Floquet theorem. This technique has been applied to RR Lyrae stars by Stellingwerf (1975a). The relaxation technique was first applied to models similar to those of Christy and a very wide either-or region was found--in agreement with Spangenberg. The survey was then repeated with improved opacity and treatment of shocks, with the results shown in Figure 5 (Figure 17 of Stellingwerf, 1975a). Both linear and nonlinear analyses were performed. The solid lines are the linear blue edges, the dashed lines are modal transition lines, and the short dashed line is the estimated red edge. In this case the either-or region is only 300 K wide for reasonable luminosities. The actual temperature at which the change of mode occurs depends

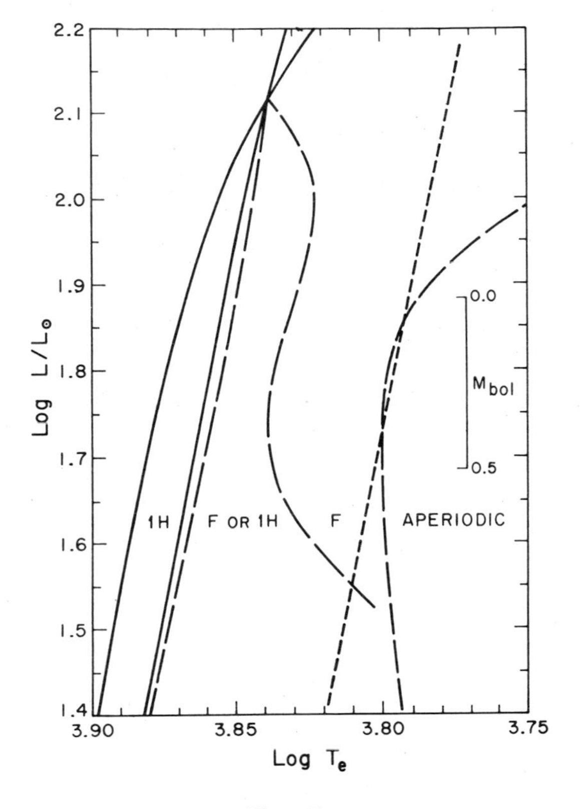

Fig. 5.

upon direction of evolution, and there is rather strong observational evidence that the resulting hysteresis effect on evolving stars is responsible for the Oosterhof dichotomy of globular clusters (van Albada and Baker 1971b, 1973). The mode switching rates depend on the rate of evolution; they appear to be on the order of 100 years for these stars. Another result seen in Figure 5 is a region toward the red in which stable mixed-mode pulsation was found, behavior very similar to that seen in the star AC And, the mixed mode Cepheids, and the dwarf Cepheids. Convection was not included in these models, so the redder models must be regarded as tenative.

AC And is the only observed RR Lyrae star showing simultaneous excitation of several radial modes. The fundamental, first overtone and second overtone modes have been detected with periods of 0.711, 0.525 and 0.421 days (Fitch and Szeidl, 1975). Stellingwerf (1975a) showed that the fundamental and first overtone periods were consistent with red horizontal branch models. Fitch and Szeidl, however, using the new second overtone data and the metal rich Population I models of Cogan (1970), found that the periods could also be fit by a 3 $M_{\odot}$ "first crossing" model with luminosity $L = 100\ L_{\odot}$ and effective temperature $T_e = 5500$ K. This is a rather cool temperature, especially since the instability strip moves to the blue for higher masses, so convection could be a factor here as well. The first-crossing hypothesis does account for the uniqueness of this object : this evolutionary phase is relatively short.

We may conclude that there exist three possibilities for the mixed-mode stars:

1) Stable, unique, nonlinear mixture of modes,
2) A process of switching between stable pure modes, or
3) A state of initial growth of pulsation, showing a mixture of quasi-linear modes.

Of these only the first can produce stable oscillations over a long period of time; the others will result in strong progressive changes in the amplitudes on a timescale of "years." This implies that continued observation could very well distinguish between these alternatives. Although the theoretical treatment of these objects is still rather sketchy, we can now make some statments about the mode-switching timescales, transition points, and overtone periods. Also, new mathematical techniques are becoming available that should make possible a more detailed comparison with observed stars.

Acknowledgements

The author's research is supported by National Science Foundation Grant GP-30927 through Columbia University. Special thanks are extended to Dr. L. B. Lucy for transmitting this communication.

References

Albada, T. S. van 1969, Bull. Am. astr. Soc., 1, 366.

Albada, T. S. van and Baker, N. H. 1971a, Astrophys. J., 169, 311.

Albada, T. S. van and Baker, N. H. 1971b, Bull. Am. astr. Soc., 3, 241.

Albada, T. S. van and Baker, N. H. 1973, Astrophys. J., 185, 477.

Baker, N. H. 1965, Kleine Veröff. Remeis-Sternw., No. 40, p122.

Baker, N. H. and Sengbusch, K. von 1969, Mitt. astr. Gesellschaft, 32, 228.

Balázs, J. 1959, Kleine Veröff. Remeis-Sternw., No. 27, p26.

Castor, J. I. 1970, Bull. Am. astr. Soc., 2, 302.

Castor, J. I. 1971, Astrophys. J., 166, 109.

Christy, R. F. 1962, Astrophys. J., 136, 887.

Christy, R. F. 1964, Rev. mod. Phys., 36, 555.

Christy, R. F. 1966, Astrophys. J., 144, 108.

Cogan, B. C. 1970, Astrophys. J., 162, 139.

Cox, J. P. 1974a, Reports on Progress in Physics, 37, 563.

Cox, J. P. 1974b, Introductory Report, Liege Astrophysical Colloquium 19.

Cox, J. P., Castor, J. I. and King, D. S. 1972, Astrophys. J., 172, 423.

Cox, J. P., King, D. S. and Stellingwerf, R. F. 1972, Astrophys. J., 171, 93.

Detre, L. and Szeidl, B. 1973, in Variable Stars in Globular Clusters and Related Systems, Ed. J. D. Fernie, D. Reidel, Dordrecht, p31.

Fitch, W. S. 1967, Astrophys. J., 148, 481.

Fitch, W. S. 1968, in Non-Periodic Phenomena in Variable Stars, Academic Press, Budapest, p287.

Fitch, W. S. 1970, Astrophys. J., 161, 669.

Fitch, W. S. and Szeidl, B. 1975, preprint.

Iben, I., Jr. 1971a, Astrophys. J., 166, 131.

Iben, I., Jr. 1971b, Astrophys. J., 168, 225.

Iben, I., Jr. 1971c, Pub. astr. Soc. Pacific, 83, 697.

Iben, I., Jr. and Huchra, J. 1970, Astrophys. J., 162, L43 .

Iben, I., Jr. and Huchra, J. 1971, Astr. and Astrophys., 14, 293.

Kluyver, M. 1936, Bull. astr. Inst. Nederland, 7, 313.

Ledoux, P. 1951, Astrophys. J., 114, 373.

Ledoux, P. 1963, in Star Evolution, Ed. L. Gratton, Academic Press, New York, p394.

Lucy, L. B. 1975, private communication.

Preston, G. W. 1964, A. Rev. Astr. Astrophys., 2, 46.

Schwarzschild, M. 1940, Harvard Circ., No. 437.

Sengbusch, K. von 1973, Mitt. astr. Gesellschaft, No. 32, p228.

Spangenberg, W. H. 1974, Ph.D. Dissertation, Univ. of Colorado.

Stellingwerf, R. F. 1974, Astrophys. J., 192, 139.

Stellingwerf, R. F. 1975a, Astrophys. J., 195, 441.

Stellingwerf, R. F. 1975b, Astrophys. J., in press.

Stellingwerf, R. F. 1975c, in preparation.

Tuggle, R. S. and Iben, I., Jr. 1972, Astrophys. J., 178, 455.

Tuggle, R. S. and Iben, I., Jr. 1973, Astrophys. J., 186, 593.

Discussion to the paper of STELLINGWERF

COX: Has Stellingwerf any further ideas about the mass of AC And?

(STELLINGWERF): The "first crossing" ($3M_{\odot}$) hypothesis is attractive for these reasons: (1) Fitch and Szeidl have obtained fits with Population I metal abundances, (2) this explains the uniqueness since the evolutionary timescale would be rather short, and (3) the mixed-mode behavior suggests similarity to the δ Scuti stars. Period fitting, however, can evidently accomodate a wide range of masses, while the other arguments are subjective in nature, so I think further data is required on this problem.

Multiple Periodic Variable Stars
IAU Colloquium No.29, Budapest, 1975

MULTIPERIODICITY IN RRs AND δ SCUTI STARS: AN OBSERVATIONAL VIEW

W.S. Fitch

Steward Observatory, University of Arizona
Tucson, Arizona 85721, U.S.A.

I. Introduction

On reviewing the observational literature of the recent past on RRs (or dwarf Cepheid or AI Velorum) and δ Scuti stars, it appeared that most present controversy is centered on three specific questions:

(1) Are δ Scuti stars really periodic, or only quasi-periodic?

(2) Are tidal modulations responsible for the slow cyclic variations observed in many of these stars, or not?

(3) Are there any real physical differences between the members of these two groups, or are they actually the same kind of object?

I would like to restrict this review primarily to these questions; and since some of our recent observational results bear directly on questions (1) and (2), and indirectly on question (3), I hope I will be forgiven for describing briefly here some of our current relevant findings.

For the most recent and comprehensive survey on δ Scuti stars, one should consult the excellent review paper by Baglin et al. (1973), which also contains a very extensive bibliography. The Annotated Catalog and Bibliography on δ Scuti Stars, by Seeds and Yanchak (1972) is quite useful for a variety of reference problems.

II. δ Scuti Variability, Periodic or Quasi-Periodic?

It has long been known that many δ Scuti stars exhibit light curves variable in phase, shape, and amplitude, so that a stable light curve is perhaps the exception to the general rule. Most presently known group members were discovered relatively recently by survey work designed for that purpose, so that only a few of these stars have been intensively observed over any extended period of time. Le Contel et al. (1974) and Valtier et al. (1974), in reporting on their photometry of HR 432, 515, 812, 8006, and 9039, first suggested and later insisted that periods in most δ Scuti stars have meaning only in a statistical sense, and that intrinsic irregularities, due probably to nonlinear coupling between convection and pulsation in the upper layers, usually dominate the variability. Smyth et al. (1975) concluded that while the primary frequency remained constant and recognizable in HR 1653 and HR 3265, in ρ Phe the frequency spectrum is an apparently continuously and rapidly varying function of the time of observation. In a private communication, Stobie has informed me that of the four stars 1 Mon, 21 Mon, ρ Phe, and σ Tuc, only 1 Mon has a stable frequency spectrum.

TABLE 1. STEWARD OBSERVATORY[1] PHOTOMETRY OF δ SCUTI STARS

Variable	Comparison	Filter	P_o(day)	Nights	Measures	Hours	Cycles	Years	Observer[3]
CC And	BD+41°120	V	0.12491	33	2230	138.2	46.1	1955-57	F
CC And	BD+41°120	V	0.12491	16	3344	92.0	30.7	1974	F[4]
14 Aur	18 Aur	b	0.08809	41	6743	160.0	75.7	1972-75	F,W
4 CVn	BD+43°2221	b	?	28	6173	149.1	?	1974	W
DQ Cep	HD199938	V	0.07886[2]	6	399	23.2	12.2[2]	1958	F
δ Del	ζ Del	b	0.15679	29	5431	114.8	30.5	1972-74	F,W
1 Mon	2 Mon	b	0.13613	5	837	18.2	5.6	1974	W
δ Sct	HR7055	b	0.19377	24	4855	111.0	23.9	1972-73	F
			Totals		30012	806.5	224.7		

1: Observations in 1972 made while Fitch was guest observar at Observatorio Nacional de Mexico, San Pedro Martir, Baja California, Mexico.

2: Primary period, but perhaps not fundamental period.

3: F = W. S. Fitch; W = W. Z. Wisniewski.

4: Observations on J.D.2442339 and 2442340 made by F. E. Brengman.

These conclusions are contradictory to my own experience. Table 1 gives a summary of our photometry on δ Scuti stars. Six of the seven stars we have observed show a well-defined and unambiguous primary period, which for simplicity I assume is the fundamental radial mode. This mode identification seems secure for CC And, δ Sct, and 1 Mon, because they also have an excited second radial overtone; but I have no evidence that the primary frequency in 14 Aur, DQ Cep, and δ Del is correctly identified. Only in the case of 4 CVn have I not yet found a stable primary frequency, and I attribute at least part of the difficulty in this case to the fact that our observations of 4 CVn were all made near full moon, so the aliasing problem with both diurnal and monthly sidelobes is very severe. It may still develop that 4 CVn is not strictly periodic, but at present I don't expect this result. I think an adequate explanation for the difference in our experience and that of the Nice group (Le Contel et al. 1974, Valtier et al. 1974) lies in the fact that δ Scuti stars with nonstable light curves usually have an extremely complex system of variability, which requires a very great amount of observing time to decode. I hope to demonstrate this conclusion here in the cases of 14 Aur, δ Sct, and CC And. For the five stars which the Nice group discussed, they had 5, 5, 4, 13, and 8 nights of observations, respectively, and the actual observing runs (and coverage of the light variation) were usually rather brief, so their failure to find regularity in the light variations is perhaps understandable.

I don't think it is presently possible to formulate any general rules concerning data adequacy for analysis of multiperiodicity, since the requirements can vary so markedly from star to star, but it is usually fairly easy to estimate when a data set is inadequate. My own prejudice is that fewer than ten long and closely spaced nights are probably not worth analysing, unless the changes in the light curve are fairly simple and easily guessed. The observations of ρ Phe in 1968 by Cousins et al. (1969), and in 1972 by Smyth et al. (1975), were made during 21.4 hours on 12 nights spanning 23 days in 1968 and during 41.4 hours on 8 nights spanning 63 days in 1972. The fractional coverage of the variability was 4.2% in 1968 and 2.7% in 1972, and the aliasing problem in both data sets is fairly severe. Having examined the data distribution and their published frequency spectra, I think that while their conclusion, that the frequency spectrum of ρ Phe is basically unstable, may be correct, it has not yet been proven. I can easily simulate the frequency spectrum behavior they found in ρ Phe by appropriate subdivisions of our data on 14 Aur, and in fact did so several years ago. Nevertheless, the primary frequency in 14 Aur is statistically a very stable entity. In Figure 1 I illustrate, with 7 typical light curves, the characteristic light variation of 14 Aur A. Normally there are about 11 1/3 cycles per day (c/d), but on some pairs of successive nights there are

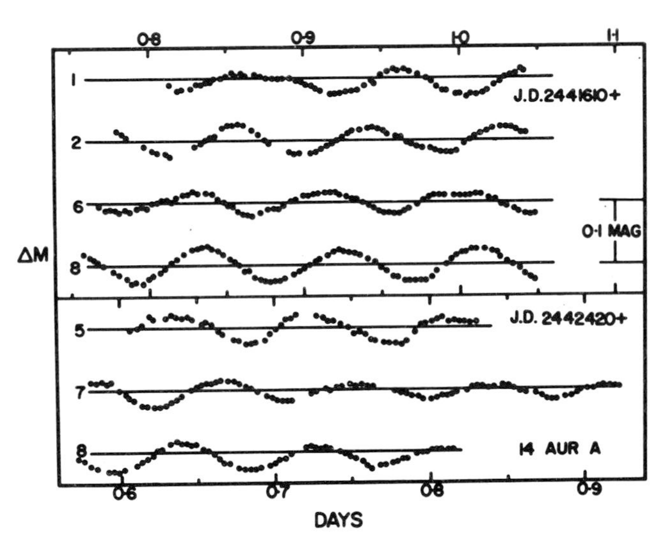

Fig. 1 - Representative observed b-magnitude light curves of 14 Aur A minus 18 Aur.

instead an integral number of c/d linking two nights. Normally there is very little cycle-to-cycle change, but sometimes the amplitude changes rapidly during one night. My present conclusions concerning 14 Aur are that the light variation is basically periodic but very complex, and I cannot pretend I yet understand it.

In Table 2 are given four different solutions approximating the blue magnitude light variation of 14 Aur A. I've not illustrated any of these

TABLE 2. b-MAGNITUDE VARIATION OF 14 AUR A: 4 REJECTED SOLUTIONS*

Name	f_i(c/d)	A_i	σ_A	ϕ_i	σ_ϕ	σ(1 obs)
$2f_L$	0.527904	0.0030	± 0.0003	0.774	± 0.016	± 8.4 mmag
f_0	11.35227	0.0168	0.0003	0.795	0.003	
$2f_L$	0.527904	0.0030	0.0003	0.776	0.015	8.3 mmag
$43f_L$	11.349936	0.0033	0.0004	0.079	0.021	
f_0	11.35227	0.0146	0.0004	0.779	0.005	
$2f_L$	0.527904	0.0032	0.0002	0.776	0.013	7.2 mmag
$43f_L$	11.349936	0.0042	0.0004	0.056	0.014	
f_0	11.35227	0.0133	0.0004	0.772	0.005	
f_{nr}	11.62581	0.0054	0.0002	0.076	0.007	
$(f_0-43f_L)/2$	0.00117	0.0048	0.0005	0.451	0.012	6.3 mmag
$f_{nr}-f_0-f_L$	0.02242	0.0028	0.0003	0.906	0.017	
$2f_L$	0.527904	0.0035	0.0002	0.792	0.010	
$43f_L$	11.349936	0.0034	0.0004	0.059	0.017	
f_0	11.35227	0.0137	0.0004	0.780	0.004	
f_0+f_L	11.616222	0.0013	0.0002	0.325	0.026	
$f_{nr}-F$	11.56180	0.0009	0.0003	0.619	0.055	
f_{nr}	11.59380	0.0025	0.0004	0.743	0.026	
$f_{nr}+F$	11.62580	0.0042	0.0004	0.028	0.015	
$f_{nr}+2F$	11.65780	0.0014	0.0004	0.681	0.047	
$f_{nr}+3F$	11.68980	0.0013	0.0003	0.396	0.040	

* $m_{14} = m_{18} - 1.507 - 2.5 \log [1 + \Sigma A_i \sin 2\pi (f_i t + \phi_i)]$

t = Hel.J.D. − 2441502.2600

1748 observations, averaged from 6743 with Δt = 0.003 day

in the observed light curves, because I don't consider any of them as acceptable representations of the star's behavior. However, while I'm not satisfied with these approximations, there are several fairly firm conclusions one can draw. First, the primary is slightly prolate, being 0.006 mag brighter when seen at velocity extrema than when seen in conjunction.

Second, there appears to be some kind of a resonance problem between the primary pulsation and the orbital motion. Third, tidal modulation such as seen in CC And and σ Sco is not present in this 3.8 day period spectroscopic binary, unless the orbital period changed suddenly and strongly between the 1968 radial velocity measures of Chevalier et al. (1968) and our own photometric measures which began in 1972. This last possibility seems much too improbable to consider further.

III. Tidal Modulation, Yes or No?

I was unhappy with my first published analysis of the light variation of CC And (Fitch 1960), because the adopted analytic representation did not adequately reproduce the observations. Later (Fitch 1967), when I found that the short and long term variability of both CC And and the β Cep star σ Sco could be understood in terms of intrinsic radial pulsation perturbed tidally by a companion in a binary orbit, I rashly suggested that whenever a short period pulsator also shows long period variations in pulsation characteristics, these long period complications are probably caused by tidal modulation. That this suggestion is not universally true has now been clearly demonstrated by Broglia and Marin (1974) for the case of Y Cam, and by myself in the cases of 14 Aur and the β Cep star 16 Lac (Fitch 1969).

Table 3 presents the orbital elements we derived from all the published velocities of 14 Aur which we could find. The period was determined by a

TABLE 3. ORBITAL ELEMENTS FOR 14 AURIGAE A

$P = 3.78857 \pm 0.00003$ days (estimated error)
$e = 0.0$
$\gamma = -9.8$ km/sec
$K = 22.5 \pm 0.7$ km/sec (m.e.)
$T_o =$ J.D.2420003.10 $\pm$ 0.02 (m.e.) at R.V. Max.
$f(M) = 0.0045 \pm 0.0004\ M_\odot$ (m.e.)

Fourier transform amplitude spectrum, giving an estimated uncertainty of 0.000002 c/d, and the rest of the elements were gotten by least-squares fits of sine curves in harmonics of the orbital period. There is no evidence for a noncircular orbit, and we do not confirm the modulation characteristic suggested by the observations of Hudson et al. (1971). Rather, given the orbital period as known, we find that the orbital phase of minimum light amplitude varies with time, just as did Broglia and Marin (1974) in the eclipsing binary Y Cam.

Table 4 presents a summary of all the material I could find on short period pulsators in close binaries (excluding pulsars, white dwarfs, etc.), arranged to emphasize the apparent correlation between long term variability and orbital eccentricity. I consider the top eleven stars as definitely established binaries, though skeptics may wish to exclude CC And from this category. The inclusion of the last three stars in this list is speculative. KP Aql (Ibanoglu and Gülmen 1974) is an eclipsing binary with an A type primary, and the published light curves suggest to me that it may also be a δ Scuti star, but this suggestion needs definite testing. I include δ Scuti and 1 Mon as binaries, because their observed characteristics are consistent with other pulsators known to be binaries, but I cannot prove my assumptions here. The assignment of the modulation characteristics to the five stars in the middle group is also speculative, by an obvious extrapolation of the apparent pattern shown by the first six stars in Table 4. GX Peg is apparently misnamed TW Lac in Table 2 of Baglin et al. (1973).

I would like to emphasize two points concerning Table 4. First, Lomb (1975) has definitely established tidal modulation of the pulsation amplitude in α Vir, so I am no longer the only one to have found this kind of behavior in a short period pulsating star. Second, I could not find any

RRs stars to include in this list, though the RRs star SZ Lyn is the primary in a wide binary with period 3.14 years (Barnes and Moffett 1975), and the RRc star RW Ari is the primary of an eclipsing binary (Wisniewski 1971).

TABLE 4. SHORT PERIOD PULSATORS IN CLOSE BINARIES

	Orbit			Pulsation			
Name*	Type	P_L(days)	e	Type	P_0(day)	Tidal Modulation	Nonradial Modes
CC And	Photom.	10.469	≈ 0.15?	δ Sct	0.12491	Yes	No
14 Aur	Sp. 1	3.7886	0.0	δ Sct	0.08809	No	Yes
Y Cam	Ecl.	3.3055	0.0	δ Sct	0.063	No	Yes?
16 Lac	Sp. 1	12.096	0.0	β Cep	0.16917	No	Yes
σ Sco	Sp. 1	33.13	0.40	β Cep	0.24684	Yes	No
α Vir	Sp. 2	4.0142	0.13	β Cep	0.1738	Yes	No?
AB Cas	Ecl.	1.3669	0.0	δ Sct	0.058	No?	Yes?
ZZ Cyg	Ecl.	0.6286	0.0	δ Sct	0.1	No?	Yes?
δ Del	Sp. 2	40.58	0.7	δ Sct	0.15679	Yes?	No?
UX Mon	Ecl.	5.9045	0.0	δ Sct	0.2	No?	Yes?
GX Peg	Sp. 1	2.3409	0.0	δ Sct	0.06	No?	Yes?
KP Aql	Ecl.	3.3675	0.0	δ Sct?			
1 Mon	Photom.	15.492?	> 0.0?	δ Sct	0.13612	Yes?	No?
δ Sct	Photom.	≈ 10?	0.0?	δ Sct	0.19377	No	Yes

* CC And, Fitch 1967, present paper; 14 Aur, present paper; Y Cam, Broglia and Marin 1974; 16 Lac, Fitch 1969; σ Sco, Fitch 1967; α Vir, Lomb 1975; AB Cas, Tempesti 1971; ZZ Cyg, Hall and Cannon 1974; δ Del, Preston, in Leung 1974, present paper; UX mon, Scaltritti 1973, Lynds 1957; GX Peg, Breger 1969, Harper 1933; KP Aql, Ibanoglu and Gülmen 1974; 1 Mon, Millis 1973, Shobbrook and Stobie 1974; δ Sct, Fath 1935, 1937, 1940, present paper.

If the suggested correlation of nonradial mode excitation and zero orbital eccentricity or of tidal modulation and nonzero eccentricity is confirmed by future work, then one possible explanation of these correlations might involve the pulsation amplitude growth rates for the δ Scuti stars. According to Chevalier (1971), the e-folding time for the fundamental radial mode amplitude is about 2×10^4 yr, or, as quoted in Baglin et al. (1973), about 10^3 yr. I don't know what the nonradial mode growth rates are, but these modes are presumably driven by energy leakage from the radial modes in the case of nonspherical symmetry, so I would expect all nonradial modes to be damped out by the continuously changing surface deformations of a primary with a close companion in an elliptical orbit. In this case only the radial modes should survive, though with phase and amplitude continuously variable in a zonal pattern over the surface of the primary. Further, of course, the integrated pulsation amplitude should be much smaller than in a single star of the same structure. If, however, the postulated close companion of a pulsating primary moves in a circular orbit, then the nonspherical deformations will appear static in the rotating frame, particularly if the primary rotates synchronously, and over a sufficient length of time the nonradial modes should grow to significant strength. This is apparently true in 16 Lac, and it may also be the explanation of the 3 nonradial modes excited in δ Sct.

Fath (1935, 1937, 1940) observed δ Sct on 63 nights in 1935, 1936, and 1938, while we have obtained a coverage of 23.9 cycles on δ Sct during 24 nights in 1972 and 1973. Fath used α Sct as a comparison star, and later found it to be variable in brightness. This introduces an unfortunate amount of noise into his measures, which are further complicated by his short (41 days maximum) observing seasons. Therefore, there is a significant uncertainty in the annual cycle counts for all but the fundamental period.

The annual sidelobes are even stronger in our own measures, where the 1972 and 1973 seasons spanned only 20 days and 14 days, respectively. It was, therefore, somewhat surprising to find that these two independent data sets, separated by 34 years, agreed on the annual cycle counts for the fundamental radial mode frequency f_o and its second harmonic $2f_o$, the second radial overtone f_2, and the strongest nonradial mode f_{n1}. Both data sets agreed on the presence of two more nonradial modes f_{n2} and f_{n3}, though f_{n3}

TABLE 5. ADOPTED SOLUTIONS FOR THE LIGHT VARIATION OF δ SCUTI*

	Fath's White Light Measures					1972-73 b-Magnitude Measures				
Name	f_i (c/d)	A_i	σ_A	ϕ_i	σ_ϕ	f_i (c/d)	A_i	σ_A	ϕ_i	σ_ϕ
f_{n3}						4.73582	0.0046	± 0.0003	0.624	± 0.010
f_0	5.16078	0.0783	± 0.0007	0.534	± 0.001	5.16070	0.0717	0.0003	0.428	0.001
f_{n2}	5.27946	0.0039	0.0007	0.215	0.028	5.27885	0.0036	0.0003	0.791	0.014
f_{n1}	5.35401	0.0176	0.0007	0.760	0.006	5.35446	0.0150	0.0003	0.732	0.003
f_2	8.59388	0.0054	0.0007	0.544	0.020	8.59355	0.0058	0.0003	0.679	0.008
$2f_0$	10.32156	0.0084	0.0007	0.915	0.013	10.32140	0.0067	0.0003	0.754	0.007
f_0+f_{n2}	10.44024	0.0025	0.0007	0.637	0.044					
$2f_{n2}$	10.55892	0.0013	0.0007	0.753	0.087					
f_0+f_{n1}	10.51479	0.0037	0.0007	0.201	0.030	10.51516	0.0025	0.0003	0.931	0.019
$2f_{n1}$	10.70802	0.0016	0.0007	0.518	0.070	10.70892	0.0016	0.0003	0.740	0.030

Fath's White Light Measures:
Comparison = α Scuti
σ (1 obs) = ± 15.6 mmag
919 measures
Δm_o (1935) = -0.048 mag
Δm_o (1936) = -0.037 mag
Δm_o (1938) = +0.014 mag

1972-73 b-Magnitude Measures:
Comparison = HR7055
σ (1 obs) = ± 7.6 mmag
1158 measures, averaged from 4855 with
Δt = 0.003 day

Δm_o = -1.189 mag

$* \; m_\delta = m_{comp.} + \Delta m_o - 2.5 \log [1 + \Sigma A_i \sin 2\pi (f_i t + \phi_i)]$
t = Hel.J.D. - 2427900.0

is very weak in Fath's data but stronger than f_{n2} in our own data. Our adopted values for f_{n2} and f_{n3} are rather sensitive to small changes in the whitening parameters for the stronger terms and may each be in error by 1 cycle/year, but this doesn't matter for the representation of these observations.

We originally assumed that the frequencies were strictly constant, and consequently experienced much difficulty in choosing the correct annual cycle counts. This occurred because while the two data sets agreed on the highest peaks for the strongest terms, they disagreed on the precise frequency values, and when we tried to force phase-locking from 1935 to 1973 we had to settle for peaks one annual sidelobe off the symmetry center of the frequency spectra. We also experienced problems in fitting to the data, and could not achieve a satisfactory analytic representation of the accurate modern photometry. We finally abandoned the assumption of constant periods and only compared the spectra of the two data sets to look for the strongest terms. Our adopted representations are compared graphically to a selection of typical observations in Figure 2, and are shown analytically in Table 5.

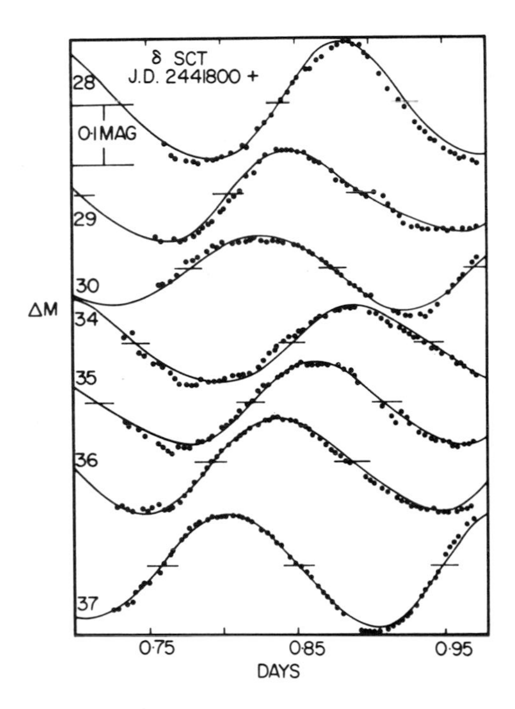

Fig. 2 - Representative observed and computed b-magnitude light curves of δ Sct minus HR7055.

We think that while these solutions could probably still be improved, they represent a reasonable approximation to the true behavior of δ Sct.

Please note, in Table 5, that the nonradial modes cluster about f_o and, in Fath's data, are apparently coupled nonlinearly to f_o, but they do not bear any simple frequency difference relation to f_o or to each other. Since this behavior is very similar to our earlier findings on 16 Lac, it prompted the suggestion that δ Sct may also be a binary with a circular orbit. Note also that the frequency pattern is not easily explained by the usual theory of rotational coupling (Ledoux 1951) or by the R- and S- mode theory of Chandrasekhar and Lebovitz (1962).

With all of the uncertainties involved, our adopted frequency changes must be regarded as highly provisional and subject to independent confirmation. If, however, they are correct, and if they represent a secular change due to evolution, then they indicate that $\delta \ln f_o = -0.000016$ in 36 yr. If we postulate evolution at constant mass M, and assume the pulsation constant $Q_o = 0.033$ day, we obtain $\delta \ln R \approx +10^{-5}$ in 36 yr, and deduce an e-folding time for the radius R of about 4×10^6 yr, with evolution toward the red side of the instability strip. There are so many possible explanations for our adopted frequency changes, including simply observational error, that little reliance should be placed on this estimate.

Because Shobbrook and Stobie (1974), in their excellent paper on 1 Mon, questioned the reality of the double cycle I found in CC And (Fitch 1967), we reobserved CC And on 16 nights in 1974. Figure 3 presents the fundamental radial mode pulsation amplitudes A_o and phases ϕ_o (obtained by fitting a sine and cosine curve to individual cycles of pulsation and then deriving the equivalent sine term $A_o \sin 2\pi (f_o t + \phi_o)$) plotted against the phases of the best 5-day period (such as is found by periodogram analysis). The phase variation in 1955-57 and the amplitude variation in 1974 both clearly show differences between even and odd cycles of the 5-day period, and they also show that the detailed modulation characteristics have changed in the interval between the two data sets. Therefore, since it is not possible to

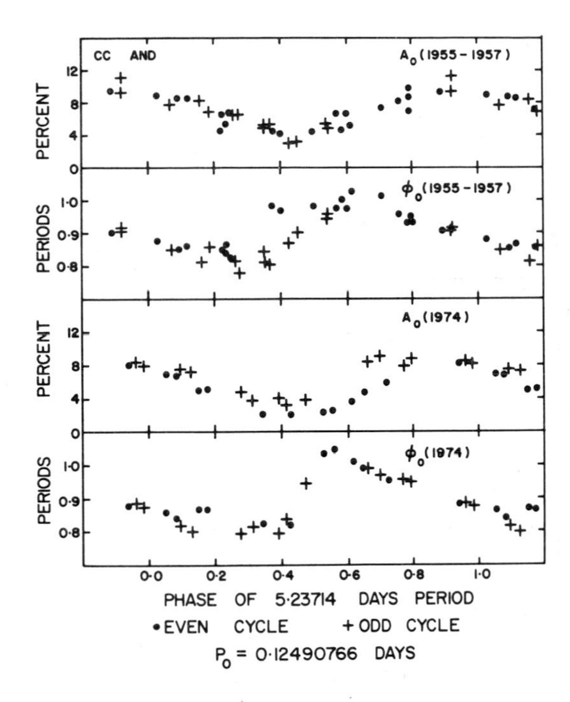

Fig. 3 - Observed phase (ϕ_o) and yellow amplitude (A_o) variation of the fundamental pulsation in CC And as functions of phase of the long period (= 5.23714 day). Even and odd cycles are differentiated.

match precisely these two sets, I derived the best value for the 10.5-day double cycle by the condition of minimum scatter in the 1955-57 data, and didn't worry about phase-locking the long period over the complete 19 yr span, though the fundamental mode is unambiguously phase-locked over this interval. Figure 4 presents the modulation characteristics of A_o and ϕ_o with phase of our adopted 10.469 day period. The smoothed A_o curves and the 1955-57 ϕ_o curve are my estimates of the real variation. The smoothed 1974 ϕ_o curve is the best 8-term analytic approximation (using a Fourier expansion including all terms through 8 f_L, where f_L = 0.09552 c/d is our adopted modulating frequency) to the smoother variation I estimated. It was necessary to obtain the analytic expansion for the 1974 phase modulation by fitting to a smoothed curve, since there is significant scatter in the

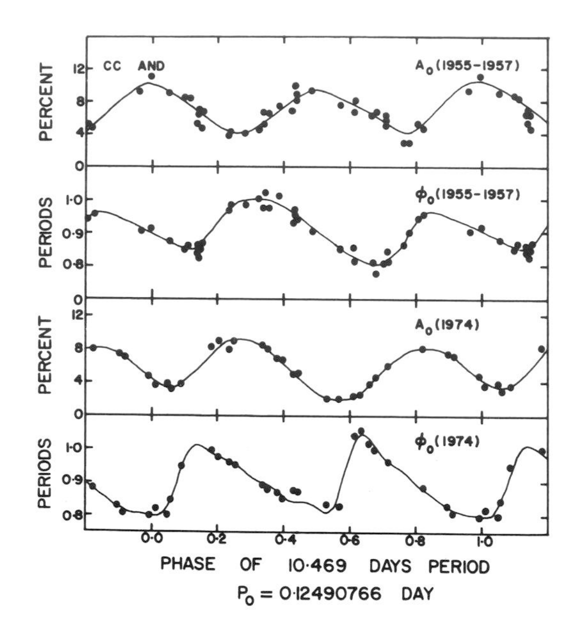

Fig. 4 - Observed (solid circles) and adopted (full line) ϕ_o and A_o variations of CC And as functions of phase of the 10.469 day period.

limited number of data points and a very high order fit is required for accurate prediction of the observed modulation. The adopted phase modulations are given in Table 6. Please note the markedly different modulation

TABLE 6. PHASE MODULATION OF CC AND*

	1955-57		1974	
i	B_i	β_i	B_i	β_i
1	0.04105	0.0073	0.01524	0.8193
2	0.06684	0.5275	0.09420	0.8621
3	0.01014	0.4913	0.00510	0.1798
4	0.01433	0.0486	0.03386	0.6772
5	0.01515	0.1390	0.00495	0.0177
6			0.01847	0.4896
7			0.00439	0.7840
8			0.01038	0.3057

* $\Delta\phi_o(t) = \Sigma B_i \sin 2\pi(if_L t + \beta_i)$
t = Hel.J.D. - 2435000.0
f_L = 0.09552 c/d

characteristics in the two data sets, which I attribute to revolution of the apse of the elliptical orbit. Both minimum and maximum amplitude are smaller in 1974 than previously, and the 1974 phase variation approaches a sawtooth pattern. If an observer only sampled the real variation on one of the branches of the 1974 phase modulation, he would by periodogram analysis derive frequency estimates for f_o ranging between 7.96 and 8.31 c/d, depending on which branch he observed. If he only sampled the real variation on two successive branches, he would conclude that the star had an unstable and continuously variable frequency spectrum. For this reason, and from my experience with δ Sct itself, I do not consider conclusions drawn from short data strings convincing, and I am extremely skeptical about the suggestion that δ Scuti stars are nonperiodic.

The second radial overtone f_2 is not directly modulated by f_L (i.e., there are no terms with frequencies $f_2 + k\ f_L$), so in our adopted representation, shown in Table 7, this mode is given by a single sine term. However,

TABLE 7. ADOPTED V-MAGNITUDE SOLUTIONS FOR CC ANDROMEDAE*

1955-57					1974				
j, k	A_{jk}	σ_A	$\emptyset_{jk}$	$\sigma_\emptyset$	j, k	A_{jk}	σ_A	$\emptyset_{jk}$	$\sigma_\emptyset$
1,-7	0.0017 ±	0.0005	0.603 ±	0.045	1,-5	0.0008 ±	0.0004	0.574 ±	0.069
1,-6	0.0010	0.0005	0.755	0.084	1,-4	0.0008	0.0004	0.731	0.071
1,-3	0.0022	0.0005	0.912	0.040	1,-3	0.0026	0.0004	0.554	0.022
1,-2	0.0151	0.0005	0.928	0.005	1,-2	0.0139	0.0004	0.492	0.004
1,-1	0.0013	0.0005	0.159	0.060	1,-1	0.0067	0.0004	0.973	0.009
1, 0	0.0676	0.0006	0.904	0.001	1, 0	0.0576	0.0004	0.901	0.001
1,+1	0.0010	0.0005	0.505	0.078	1,+1	0.0068	0.0004	0.763	0.009
1,+2	0.0134	0.0005	0.884	0.006	1,+2	0.0152	0.0004	0.288	0.004
1,+3	0.0037	0.0005	0.814	0.024	1,+3	0.0015	0.0004	0.501	0.040
1,+6	0.0026	0.0005	0.906	0.031	1,+4	0.0026	0.0004	0.190	0.022
1,+7	0.0029	0.0005	0.226	0.027	1,+5	0.0029	0.0004	0.073	0.020
2,-2	0.0044	0.0005	0.782	0.016	2,-2	0.0048	0.0003	0.329	0.011
2, 0	0.0087	0.0005	0.676	0.008	2, 0	0.0066	0.0003	0.620	0.008
2,+2	0.0044	0.0004	0.617	0.016	2,+2	0.0037	0.0003	0.070	0.015
f_2	0.0078	0.0005	0.638	0.009	f_2	0.0047	0.0003	0.636	0.011

1955-57	1974
$\Delta m_o = +0.263$ mag	$\Delta m_o = +0.264$ mag
σ (1 obs) = ± 10.6 mmag	σ (1 obs) = ± 7.4 mmag
1041 measures, averaged from 2230 with $\Delta t = 0.0035$ day	865 measures, averaged from 3344 with $\Delta t = 0.0035$ day

* $m_{CC} = m_{comp} + \Delta m_o - 2.5 \log \{ 1 + \Sigma A_{jk} \sin 2\pi [(jf_0 + kf_L)t + j\Delta\emptyset_o(t) + \emptyset_{jk}] + A_2 \sin 2\pi (f_2 t + \emptyset_2) \}$

$f_0 = 8.005914$, $f_2 = 13.346146$, $f_L = 0.09552$ c/d. t = Hel.J.D. - 2435000.0

the systematic nature of the residuals from our adopted solution, shown for the 1974 data in Figure 5, strongly suggests that most of the remaining fitting errors above the noise level are due to the neglect of modulation on f_2. Time limitations in the preparation of this paper prevented following up this point, but I plan to see whether one can represent the modulation of f_2 through coupling to f_o (i.e. with terms having phases of the form $jf_ot + j\Delta\phi_o(t) + kf_Lt + mf_2t$).

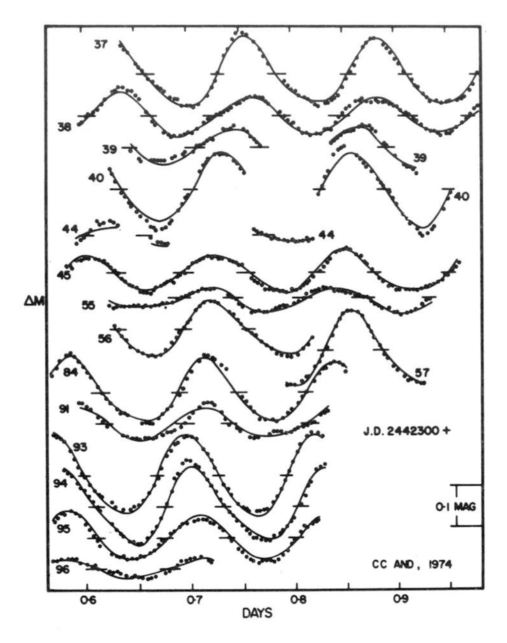

Fig. 5 - Observed and computed yellow magnitude variation of CC And minus BD + 41°120 in 1974.

In concluding this section, I should like to emphasize the following points:

(1) At present, it appears that a short period pulsator in a close binary will exhibit tidal modulation of radial pulsation modes if the orbit is elliptical, and will show both radial and nonradial mode excitation if the orbit is circular. However, in this last case the nonradial mode

frequency spacing is not at a simple constant frequency difference.

(2) In at least the case of CC And, the accurate representation of the observed light variation requires a complex and highly nonlinear modulation of amplitude and phase, separately. It is not possible to achieve a good approximation to the observed variation using only a simple amplitude modulation function involving just the sum of a set of sine terms each having a constant frequency and phase. This implies that the observed variation is actually an integral over the apparent disk of areas which are moving instantaneously with different phases and amplitudes, but with the same (fundamental) frequency. This seems to me physically plausible, since the radial mode variations in the deep interior will be little affected by tides, whereas both the phase and amplitude of the emergent wave are sensitive and highly nonlinear functions of the detailed structure of the HeII (and possibly also H and HeI) ionization zone, which must itself be zonally perturbed by any close companion.

(3) Aside from possible slow secular or very long period changes in frequencies and relative mode strength, the variations of δ Sct are strictly periodic. The variations of CC And are also strictly periodic (or very slowly changing in secular manner), and they require at most the fundamental and second radial overtone period, the orbital period, and the apse precessional period to explain them.

(4) Ordinary Fourier transform calculations, whether with data whitening in the time domain (Wehlau and Leung 1967) or with spectrum whitening in the frequency domain (Gray and Desikachary 1973), are not adequate tools for dealing with the complexities of variation displayed in such stars as CC And. Much better analytical methods, allowing for nonlinear phase and amplitude modulations, are urgently needed. Realistic models for such stars as CC And

will also require provision for the lag between the instantaneous amplitude and phase, and the instantaneous growth rates of the pulsation, all of which I think vary in a continuous manner over the stellar surface.

IV. Are RRs and δ Scuti Stars Physically Distinct?

Eggen (1956) first suggested that there was a real difference between the low amplitude δ Scuti stars and the higher amplitude RRs stars, though both groups show spectral types A to F and have periods less than 0.25 day (< 0.21 day, according to the General Catalog of Variable Stars, though VX Hya does not conform to this definition). More recently Eggen (1970), Breger and Bregman (1975), Chevalier (1972), and Baglin et al. (1973) have all argued, from various points of view, that there is no meaningful difference between the two types. Petersen and Jϕrgensen (1972) agreed with the prevailing view that δ Scuti stars are main-sequence and early post-main-sequence Pop. I stars with masses $1.5 \leq M/M_{\odot} \leq 2.5$, while concluding that the RRs stars are post-red-giant Pop. II objects with $M < M_{\odot}$. In a recent communication, Petersen has informed me that he now thinks all RRs stars, including SX Phe, may have $M > M_{\odot}$, in agreement with the δ Scuti stars. Breger (1975), from a Wesselink analysis based on published velocities and new uvbyβ photometry, has derived a radius $R = 3.0\ R_{\odot}$ and mass $M = 2.9\ M_{\odot}$ for the RRs star AD CMi. The mass determination in this case depends on log g measures, which I think are generally unreliable. If instead one adopts Breger's value for the radius, and a fundamental radial mode pulsation constant $Q_o = 0.033$ day, a more reasonable mass $M = 1.9\ M_{\odot}$ follows. In either case, Breger's evidence for a Pop. I mass in AD CMi seems to me very convincing.

Walraven's (1955) classic paper on SX Phe and AI Vel pioneered all more recent investigations of multimode excitation of Cepheid strip stars. While trying to develop suitable analytic procedures for computer simulation of the observed variations in VX Hya (Fitch 1966), I found it is only necessary to use a doubly-harmonic Fourier expansion to reproduce the light variation in that doubly-periodic star. Because I have since verified that the method also works very well on the published photometry (and velocities) of the double radial mode variables SX Phe and AI Vel (Walraven 1955) and VZ Cnc (Fitch 1955), I assume it will also work on all other such stars which lack additional complications such as, for example, tidal modulation. Very recently, Fitch and Szeidl (1976) found that the peculiar RR Lyrae star AC And has the fundamental and first and second radial overtone modes all excited and nonlinearly coupled together, so that the analytic representation of the light variation requires a triply-periodic harmonic expansion. To obtain even a rough approximation to the observed variation requires a fifth order expansion containing 115 different frequencies generated from the three incommensurable radial mode periods.

Table 8 is an extract from Table 3 of our paper on AC And, listing those observed radial mode periods and period ratios for δ Scuti, RRs, and

TABLE 8. SHORT PERIOD MULTIMODE VARIABLES IN THE CEPHEID STRIP

Name*	Type	P_0(day)	P_1(day)	P_2(day)	P_1/P_0	P_2/P_1	P_2/P_0	Log $\rho/\rho_\odot$	$M/M_\odot$	$R/R_\odot$
SX Phe	RRs	0.05496	0.04277		0.7782			-0.50		
CY Aqr	RRs	0.06104	0.04543:		0.7443:			-0.52:		
AE UMa	RRS	0.08602	0.06653		0.7734			-0.87		
RV Ari	RRs	0.09313	0.07195		0.7726			-0.94		
21 Mon	δ Sct	0.09991	0.07500		0.7507			-0.95		
BP Peg	RRs	0.10954	0.08451		0.7715			-1.07		
AI Vel	RRs	0.11157	0.08621		0.7727			-1.09		
CC And	δ Sct	0.12491		0.07493			0.5999	-1.15		
1 Mon	δ Sct	0.13612		0.08261			0.6069	-1.23		
V703 Sco	RRs	0.14996	0.11522		0.7683			-1.33		
δ Sct	δ Sct	0.19377		0.11636			0.6005	-1.52	≈ 1.9	≈ 4.0
VX Hya	RRs	0.22339	0.17272		0.7732			-1.68		
VZ Cnc	RRs		0.17836	0.14280		0.8006		-1.70		
AC And	RRab	0.71124	0.52513	0.42107	0.7383	0.8018	0.5920	-2.61	3.1	10.7

: P_1 for CY Aqr is uncertain

* SX Phe, Walraven 1955; CY Aqr, Fitch 1973; AE UMa, Szeidl 1974; RV Ari, Detre 1956; 21 Mon, Gupta 1973; BP Peg, Broglia 1959; AI Vel, Walraven 1955; CC And, present paper; 1 Mon, Shobbrook and Stobie 1974; V703 Sco, Oosterhoff 1966; δ Sct, present paper; VX Hya, Fitch 1966; VZ Cnc, Guman 1955, Fitch 1955; AC And, Fitch and Szeidl 1976.

RR Lyr stars which I consider to be reliably determined (the overtone period in CY Aqr is still open to question). The tabulated densities are calculated from the observed periods by fitting formulae developed from Cogan's (1970) published pulsation models. These formulae also permit direct calculation of mass M and radius R in favorable cases, and the observed P_o and P_2 periods of AC And lead to the mass and radius we adopted, but they do not permit such a calculation for the shorter period variables here. The mass and radius of δ Sct were adopted from the density (inferred from the observed periods P_o and P_2), the trigonometric parallax (Jenkins 1952), mean apparent magnitude (Hoffleit 1964), surface temperature (Bessell 1969), and metal rich model star evolutionary tracks (Robertson 1971, 1975). Evolution theory, pulsation theory, and observation all agree if δ Sct and AC And both have $Z \approx 0.044$ and are on their first left-to-right crossing of the instability strip. Pulsation theory, as represented by the models of Cogan (1970) and of Petersen and Jϕrgensen (1972) agrees with observation, as represented by the observed ΔS values (Preston 1959) and periods for these stars. But since Chevalier (1972) has challenged the model calculations of Cogan and of Petersen and Jϕrgensen, and since Petersen has apparently now changed his stand regarding the nature of the RRs stars, I cannot predict what the final outcome of the argument will be.

Please note that of the 14 stars listed in Table 8, only VZ Cnc does not have an excited fundamental radial mode, that 5 of these stars have firmly established excitation of the second radial overtone, and that the three δ Scuti stars CC And, 1 Mon, and δ Sct have a weakly excited second overtone and quiescent first overtone. I conclude that there are no obvious differences between the radial mode periods of the RRs and δ Scuti stars, except those due to the (perhaps debatable) dependence on heavy element composition Z.

Breger (1970) argued that A-F pulsators and Am stars are mutually exclusive groups, and that perhaps metallicism inhibits pulsation. This view has been pursued further by Baglin (1972), by Vauclair et al. (1974), and by others, and will be described by Baglin at this meeting. The crux of the argument seems to be that in slowly rotating single stars, element separation by diffusion can lead to a He deficiency in the He II ionization region, so that such stars lack the capability for self-excited oscillations, whereas stars maintained with a homogeneous envelope by rotationally driven mixing currents will start to pulsate as they pass through the instability strip. I've nothing to add to this discussion, but I should like to offer one other simple suggestion for consideration.

The correlation of increasing pulsation amplitude with increasing period and luminosity and decreasing surface temperature, when moving up the center of the instability strip from the main sequence, is well known, and reasonably well explained by model calculations. If we assume that the RRs and δ Scuti stars are all Pop. I or disk population objects with $M > M_{\odot}$, and if we compare, from Tables 3 and 8 (and SZ Lyn), stars with the same fundamental period, perhaps the amplitude differences between RRs and δ Scuti stars merely reflect the presence or absence of complications caused by close companions. That is, comparing the RRs stars VX Hya or VZ Cnc, SZ Lyn, RV Ari, and AE UMa with the δ Scuti stars δ Sct, CC And, 21 Mon, and 14 Aur A, respectively, it may well be that the larger amplitudes in the first group are those normal to essentially single stars, while the smaller amplitudes of the second group result from partial damping of the radial modes by the nonspherical deformations caused by close companions. If so, then here is one more mechanism for inhibiting pulsation in the instability strip. I am not now foolish enough to suggest that all δ Scuti stars with long period complications are members of binary systems, but I do think it possible that there are many more of them in close binaries than we presently recognize. In any case, it seems unlikely that any one mechanism will explain the great diversity of characteristics displayed by these very interesting short period variables.

In conclusion, I would venture to suggest that in any future discussion of these stars, we could greatly profit by closer collaboration with our colleagues in IAU Commissions 26, 30, and, especially, 42.

The new observations of δ Sct and CC And herein described will be available from the archives in London and Odessa as files IAU(27).RAS-42 and 43, respectively.

I'm greatly indebted to Dr. Arcadio Poveda and the staff of the Observatorio Nacional de Mexico, at San Pedro Martir, for the observing time and kind hospitality and assistance we enjoyed there. Our data analysis was performed on the CDC 6400 at the University of Arizona Computer Center. Our own work was supported in part by National Science Foundation grant GP-38739.

References

Baglin, A. 1972, Astr. Astrophys., 19, 45.

Baglin, A., Breger, M., Chaevalier, C., Hauck, B., Le Contel, J.-M., Sareyan, J.-P., Valtier, J.-C. 1973, Astr. Astrophys., 23, 221.

Barnes, T. G., Moffett, T. J. 1975, A.J., 80, 48.

Bessell, M. S. 1969, Ap.J. Suppl., 18, 167.

Breger, M. 1969, A.J., 74, 166.

Breger, M. 1970, Ap.J., 162, 597.

Breger, M. 1975, in press.

Breger, M., Bregman, J. N. 1975, in press.

Broglia, P. 1959, Nuova serie (Milano): Contr. Oss. astr. Milano-Merate, No. 142.

Broglia, P., Marin, F. 1974, Astr. Astrophys., 34, 89.

Chandrasekhar, S., Lebovitz, N. R. 1962, Ap.J., 136, 1105.

Chevalier, C. 1971, Astr. Astrophys., 14, 24.

Chevalier, C. 1972, in Stellar Ages, Proc. IAU. Colloquium, No. 17 (Paris-Meudon Obs.).

Chevalier, C., Perrin, M. N., Le Contel, J.-M. 1968, Astrophys. Letters, 2, 175.

Cogan, B. C. 1970, Ap.J., 162, 139.

Cousins, A. W. J., Lagerwey, H. C., Shillington, F. A., Stobie, R. S. 1969, (Capetown): Mon. Notes astr. Soc. Sth. Afr., 28, 25.

Detre, L. 1956, (Budapest): Mitt. Sternw. ungar. Akad. Wiss., No. 40.

Eggen, O. J. 1956, Pub. A. S. P., 68, 238.

Eggen, O. J. 1970, Pub. A. S. P., 82, 274.

Fath, E. A. 1935, Lick Obs. Bull., 17, 175.

Fath, E. A. 1937, Lick Obs. Bull., 18, 77.

Fath, E. A. 1940, Lick Obs. Bull., 19, 77.

Fitch, W. S. 1955, Ap.J., 121, 690.

Fitch, W. S. 1960, Ap.J., 132, 701.

Fitch, W. S. 1966, Ap.J., 143, 852.

Fitch, W. S. 1967, Ap.J., 148, 481.

Fitch, W. S. 1969, Ap.J., 158, 269.

Fitch, W. S. 1973, Astr. Astrophys., 27, 161.

Fitch, W. S., Szeidl, B. 1976, Ap.J., Feb. 1, 1976, in press.

Gray, D. F., Desikachary, K. 1973, Ap.J., 181, 523.

Guman, I. 1955, (Budapest): Mitt. Sternw. ungar. Akad. Wiss., No. 36.

Gupta, S. K. 1973, Observatory, 93, 192.

Hall, D. S., Cannon, R. O. 1974, Acta astr., 24, 79.

Harper, W. E. 1933, Publ. Dom. astrophys. Obs., Victoria, 6, 203.

Hoffleit, D. 1964, Catalog of Bright Stars (New Haven, Conn.: Yale Univ. Obs.)

Hudson, K. I., Chiu, H. Y., Maran, S. P., Stuart, F. E., Vokac, P. R. 1971, Ap.J., 165, 573.

Ibanoglu, C., Gulmen, O. 1974, Astr. Astrophys., 35, 487.

Jenkins, L. F. 1952, General Catalog of Trigonometric Stellar Parallaxes (New Haven, Conn.: Yale Univ. Obs.)

Le Contel, J.-M., Valtier, J.-C., Sareyan, J.-P., Baglin, A., Zribi, G. 1974, Astr. Astrophys. Suppl., 15, 115.

Ledoux, P. 1951, Ap.J., 114, 373.

Leung, K.-C. 1974, A.J., 79, 626.

Lomb, N. R. 1975, Doctoral Thesis, School of Physics, Univ. Sydney (Sydney, N. S. W.)

Lynds, C. R. 1957, Ap.J., 126, 69.

Millis R. L. 1973, Pub. A. S. P., 85, 410.

Oosterhoff, P. Th. 1966, BAN, 18, 140.

Petersen, J. O., Jørgensen, H. E. 1972, Astr. Astrophys., 17, 367.

Preston, G. W. 1959, Ap.J., 130, 507.

Robertson, J. W. 1971, Ap.J., 170, 353.

Robertson, J. W. 1975, private communication.

Scaltritti, F. 1973, (Roma): Mem. Soc. astr. ital., 44, 387.

Seeds, M. A., Yanchak, G. A. 1972, The Delta Scuti Stars: An Annotated Catalog and Bibliography (Bartol Research Foundation, The Franklin Institute, Swarthmore, Penna.).

Shobbrook, R. R., Stobie, R. S. 1974, M. N. R. A. S., 169, 643.

Smyth, M. J., Shobbrook, R. R., Stobie, R. S. 1975, M. N. R. A. S., 171, 143.

Szeidl, B. 1974, (Konkoly-Budapest): Comm. 27 IAU. Inf. Bull. var. Stars, No. 903.

Tempesti, P. 1971, Comm. 27 IAU Inf. Bull. var. Stars, No. 596.

Valtier, J.-C., Sareyan, J.-P., Le Contel, J.-M., Zribi, G. 1974, Astr. Astrophys. Suppl., 18, 235.

Vauclair, G., Vauclair, S., Pamjalnikh, A. 1974, Astr. Astrophys., 31, 63.

Walraven, Th. 1955, BAN, 12, 223.

Wehlau, W., Leung, K.-C. 1964, Ap.J., 139, 843.

Wisniewski, W. Z. 1971, Acta astr., 21, 307.

Discussion to the paper of FITCH

DESIKACHARY: Do you identify the nonradial oscillations you propose for cases like δ Sct with any specific modes such as gravity and f modes?

FITCH: No. That problem I leave to the theoreticians.

DESIKACHARY: Would you like to explain the clustering of the nonradial modes around the fundamental?

FITCH: In δ Sct, there is evidence that the nonradial modes are coupled to the fundamental, and I assume they derive their excitation by energy leakage from f_o. If this is correct, it seems to me natural that the nonradial modes observed will be those relatively close to the strongest radial mode excited.

BAGLIN: If nonradial modes are the dominant ones, this could have some consequences on the observational parameters. The fact that the different parts of the surface do not pulsate in phase would perturb the amplitudes and the relation between the measured rotational velocity and light variations. The relation between ΔV_R and Δm for high amplitude δ Scuti stars seems to agree with the Cepheids, which are radial pulsators. For example, the β CMa stars follow a very different relation - Lucy has shown in this meeting that a mixture of nonradial modes could look more like macroturbulence than like pulsation. The light curves analysed in this paper correspond to evolved stars (class IV and III) and slowly rotating stars ($V \simeq$ 30-60 km/sec) - they differ very much from the classical ordinary variables close to the main sequence, as, for example HR 8006.

LE CONTEL: I want to make a suggestion for the general discussion on these δ Scuti stars. I think we have to distinguish at least 3 subgroups: "dwarf" δ Scuti stars (rapid rotators, small amplitudes, and complicated variations), for example HR 8006; giants (rapid rotators, low amplitude variations) for example HR 515, HR 432; and a third group you present, which are almost all binaries or slow rotators (14 Aur).

In my opinion, if all these stars have the same mechanism at work from the point of view of internal structure, their observed differences are probably real due to the fact that we only observe superficial layers. So you are probably right in the third case where tidal interaction can produce the main effect, but in the other two cases we have to look for other physical phenomena which could be at work in the atmosphere. This is why we suggest examination of the coupling between pulsation and convection.

GEYER: I do not agree with you that RRs stars and δ Scuti stars are closely connected. There are two RRs stars observed in globular clusters, the membership of which has up to now not completely been excluded nor established. V65 in ω Centauri is, according to the Greenwich proper motion investigation perhaps a non-member. On the other hand, it falls so well within the RR Lyrae gap of this cluster and shows the same UV excess as these RR Lyrae stars, that it might as well be a cluster member.

FITCH: Personally, I think there must be a continuum of properties for the RRs stars, with masses and compositions ranging from those of the δ Scuti stars to those of the extreme halo population.

Multiple Periodic Variable Stars
IAU Colloquium No.29, Budapest, 1975

δ SCUTI AND RRs TYPE VARIABLES : THEORETICAL ASPECTS

J. Otzen Petersen
Copenhagen University Observatory
Øster Voldgade 3, DK-1350 Copenhagen, K Denmark

Abstract

Theoretical problems in connection with δ Sct and RRs type variables are reviewed. All evidence shows that δ Sct stars are normal main-sequence or early post main-sequence stars. Results of linear stability analyses of δ Sct models agree well with observations, the presence of non-variables in the strip being understandable in terms of helium diffusion. Excitation of non-radial modes may also occur; comparison of observed multiple periodicities with the theoretically derived period patterns for radial and non-radial modes is discussed as a means to distinguish between these possibilities.

It is not known whether RRs variables are of the same nature as δ Sct stars or in a late evolutionary state with low mass. RRs variables pulsate in radial modes, and recent linear stability analyses seem to agree well with observations for both normal and low masses. The reason for the difference in amplitude between RRs and δ Sct stars is not known. Non-linear investigations may provide important information in the near future. The observed period ratios of RRs variables indicate for most of the double mode pulsators a slight deficiency in heavy elements and a normal mass; but more detailed investigations of this problem are needed.

1. Introduction

Each group of variable stars raises questions of two types to theoreticians: (i) Can we understand the observed light- and velocity-curves by means of pulsation theory? (ii) Where do these stars fit into stellar evolution theory, and is there agreement on their properties inferred from pulsation and from evolution theory? This comparison provides very important tests of stellar evolution theory.

The δ Scuti stars are variables (i) situated in the Cepheid instability strip in the HR diagram above the zero-age line and below the RR Lyrae region (ii) which have relatively small pulsation amplitudes ($\Delta m_V < 0.3$), and (iii) periods less than 0.25 days. All evidence, observational as well as theoretical, seems to confirm the simplest possible assumption: δ Sct stars are Population I A or F stars undergoing normal evolution in the main-sequence stage or in the shell hydrogen burning phase immediately following the main-sequence stage.

RRs variables are distinguished from the δ Sct group by their larger amplitudes ($0.3 < \Delta m_V < 0.8$), but in most other respects these two types of variables have very similar properties. Parallax determinations of the RRs variables AI Vel and SX Phe indicate that these stars are placed below the Population I zero-age main-sequence (See e.g. Bessell, 1969b; Fitch, 1970); and RRs stars have often been suggested to be in a late evolutionary phase (Kippenhahn, 1965; McNamara, 1965; Bessell, 1969b; Petersen and Jørgensen, 1972; Dziembowski and Kozlowski, 1974; Percy, 1975). However the evidence, both observational and theoretical, for an intrinsic, evolutionary difference between δ Sct and RRs stars is rather weak (e.g. Baglin et al., 1973; McNamara and Langford, 1974). At present it seems possible that these two groups of variables are intrinsically similar, the only difference being that a few of the RRs stars belong to the old disk population or even the halo population of metal deficient stars. We return to these problems in more detail later.

Also detailed analysis of the observed complicated light- and velocity curves seems to raise quite different problems for δ Sct and RRs variables. Since δ Sct and RRs stars populate the lower part of the Cepheid instability strip, we expect them to oscillate in simple radial modes excited primarily in the second ionization zone of helium, as other well-behaved Cepheids. Already Kippenhahn and Hoffmeister (Kippenhahn, 1965) found a weak instability in main-sequence models corresponding to δ Sct stars and a more pronounced one for low mass models which could be interpreted as RRs variables. More detailed investigations by Castor (1970), Chevalier (1971), Cox et al. (1973), Dziembowski and Kozlowski (1974), and Percy (1975) have confirmed these theoretical results, which will be discussed in more detail in the following sections.

However, the light- and velocity-curves of δ Sct and RRs variables manifestly do not look like those of well-behaved classical Cepheids or RR Lyrae variables. It has been particularly difficult to interpret detailed observations of δ Sct stars. At present it is not clear whether δ Sct pulsations are basically simple radial oscillations or not. Dziembowski (1974) has shown that also non-radial modes are excited in δ Sct models, and perhaps the light-curves are more easily understood in terms of non-radial oscillations. Also other complicating effects may play a role for δ Sct variables (Fitch, 1970; Le Contel et al., 1974; Warner, 1974; Fitch, this volume).

Besides, the theoretician has to face the problem of the presence of non-variables in the δ Sct instability strip. Why are about 70 per cent of the stars in the observed instability strip constant to within 0.01 mag? Baglin (1972) showed that the correlation observed in main-sequence stars between pulsation, metallicity, and rotation can be understood in terms of a competition between diffusion processes, that tend to separate the elements in radiative regions of stars, and mixing due to meridional circulation. In very slowly rotating stars with negligible circulation, downwards diffusion in the outermost layers removes helium from the driving He^+ ionization zone, while some "metals" tend to move to the surface (radiative force). Since the main excitation mechanism can not operate in such Am stars they will not pulsate. In fast rotators, on the other hand, meridional circulation provides efficient mixing. Thus fast rotators are able to pulsate, but not to develop metallicism. Breger (1972) suggested a more elaborate scheme for the properties of stars in the δ Scuti region. The extensive non-adiabatic calculations by Cox et al. (1973) have confirmed that a helium content of less than about 10 per cent in the driving He^+ ionization zone will result in pulsational stability; and Vauclair et al. (1974) have studied the change in time of the helium profile of two typical δ Sct envelope models by detailed computation of the diffusion process. The separation takes less than 1 per cent of the main-sequence life time, so non-rotating main-sequence stars in the instability strip will not pulsate.

Many RRs variables show amplitude modulations that can be understood as due to simultaneous excitation of two radial modes. For 7 RRs stars Fitch (1970) gave two accurate periods, and showed that in 6 cases the fundamental radial mode and the first over-tone is excited, while 1 star oscillates in the first and second over-tone. Theoretical problems in connection with this type of model behaviour has been greatly clarified by the results of Stellingwerf (1974, 1975b) who studied models of RR Lyrae stars and double-mode Cepheids of periods 2-4 days. A similar investigation of models in the lower part of the Cepheid instability strip will probably provide valuable new information.

At present the reason for the difference in pulsation amplitudes between typical δ Sct and RRs variables is not understood. The properties of these stars seem to be very similar, except that RRs stars are slow rotators, while the mean projected rotational velocity is about 90 km/s for low amplitude δ Sct stars. Also, a small fraction of the RRs variables belong to the metal deficient old disk population, while δ Sct stars seem to belong to young Population I. It has been pointed out, however, that there is a continuous variation in properties from high amplitude ($\Delta m_V \cong 0.3$) δ Sct stars to the RRs variables of similar amplitudes (Eggen, 1970; Baglin et al., 1973). And it seems possible that the delicate balance between the complicated processes that tend to mix and those that separate the stellar matter in the very thin layers that excite or damp the oscillations, in some cases will produce a particularly strong excitation.

In the following we first discuss the δ Scuti group, emphasizing the problem of the excitation mechanism, and then the RRs type variables whose nature and evolutionary state have been widely debated.

2. δ Scuti Stars

2.1 Evolutionary stage

In order to discuss the structure and evolution of δ Sct stars several authors have performed determinations of effective temperature, T_e, gravity, g, luminosity, L or M_{bol}, mass, M, and pulsation constant, Q (cf. the review article by Baglin *et al.*, 1973). The stellar atmosphere parameters T_e and g can be obtained from photometric and spectrophotometric measurements (Bessell, 1969a; Breger and Kuhi, 1970; Dickens and Penny, 1971).

When an independent absolute magnitude is available from trigonometric parallax or cluster membership, or the like, an empirical mass value and pulsation constant can be computed from

$$\log M/M_\odot = 12.502 + \log g - 0.4\, M_{bol} - 4 \log T_e \qquad (1)$$

$$\log Q_{obs} = -6.454 + \log \Pi + 0.5 \log g + 0.1\, M_{bol} + \log T_e \qquad (2)$$

where Π is the period in days. Typical observational uncertainties in favourable cases are $\Delta \log T_e \cong 0.01$, $\Delta \log g \cong 0.3$, $\Delta M_{bol} \cong 0.5$, and $\Delta \log \Pi \cong 0.04$. (1) and (2) then give error estimates $\Delta \log M/M_\odot \cong 0.37$ and $\Delta \log Q \cong 0.16$; i.e. present uncertainties in M and Q derived directly from photometric and spectroscopic stellar atmosphere parameters combined with luminosities from parallaxes etc. are factors of at least 2.5 and 1.4 respectively. Hence, by such investigations it is not possible to decide whether masses of δ Sct stars are equal to the evolution masses, derived by comparison of their position in the HR diagram with standard evolutionary tracks for a typical Population I composition, or perhaps a factor 2 smaller. Nor can it be decided which mode is excited, because the difference between successive modes is at most 25 per cent. However, the masses and pulsation constants determined in this way by Breger and Kuhi (1970) and Dickens and Penny (1971) are in agreement with (the evolutionary) masses $M = 1.5 - 2.5\ M_{\odot}$ (cf. Fig.1) and pulsation in the fundamental mode or an over-tone of low order.

If we rely on the assumption that δ Sct stars are Population I stars of normal type undergoing standard evolution, we can use standard calibrations of e.g. uvby, β photometry to derive M_{bol} and $\log g$ to a higher accuracy than just mentioned (Strömgren, 1966; Crawford, 1970; Petersen and Jørgensen, 1972; Breger, 1972, 1974, 1975a). Estimating $\Delta M_{bol} \cong 0.2$, $\Delta \log g \cong 0.1$, $\Delta \log T_e \cong 0.01$, and $\Delta \log \Pi \cong 0.04$ we now find $\Delta \log M/M_{\odot} \cong 0.12$ and $\Delta \log Q \cong 0.07$. Thus in favourable cases it may be possible to distinguish pulsation in the fundamental mode from over-tone pulsations. The results by Breger (1972, 1975a) and Petersen and Jørgensen (1972) indicate that both fundamental mode and over-tone oscillations occur among δ Scuti stars. Masses found in such investigations are $M = 1.5 - 2.5\ M_{\odot}$ as expected.

Table 1. Ages for Population I stars of composition (X,Z) = (0.70,0.03) in the instability strip in the main-sequence or early post main-sequence phases. Min. (max.) refers to the left (right) endpoint of the relevant section of the evolutionary track (cf. Fig.1). The fast hydrogen exhaustion phase has been neglected.

	Main-sequence			Post main-sequence		
		Age (unit: 10^6y)			Age (unit: 10^6y)	
$M/M_{\odot}$	M_{bol}	Min.	Max.	M_{bol}	Min.	Max.
1.6	2.6	0	580			
1.8	2.0	0	1020	1.6	1030	1130
2.0	1.5	495	728	1.1	743	817
2.5				0.3	430	434

Thus δ Sct stars seem to be Population I, young disk stars evolving through the instability strip in the main-sequence phase or in the much faster shell hydrogen burning phase immediately following the main-sequence. From age calibrations (Hejlesen, 1975) their age and the time interval they spend in the instability strip can be estimated. A few typical data for $(X,Z) = (0.70,0.03)$ are given in Table 1.

2.2 Excitation Mechanism, Multiple Periodicities

Linear stability investigations of δ Scuti models by Kippenhahn and Hoffmeister (Kippenhahn, 1965), Castor (1970), Chevalier (1971), Cox *et al.* (1973), Pamjatnikh (1974), Dziembowski (1974, 1975), and Stellingwerf (1975c) have shown that inside the observed instability strip models that have the standard helium content $Y \cong 0.30$ are unstable in the fundamental mode and the first few over-tones. Cox *et al.* investigated the position of the fundamental and the first over-tone blue edges for several compositions, varying Y from 0.37 to 0.00. Their results are shown in Fig.1. For $Y = 0.28$ the calculated blue edge for the second over-tone agrees well with the observed high temperature boundary of the instability strip. For higher Y the calculated blue edges become hotter, while for $Y = 0$ no instability is found in the observed strip. Of course, this type of variation with Y is to be expected, since the second ionization zone of helium provides the driving of the oscillations.

From the results just mentioned one should expect all stars of standard Population I composition in the instability strip to oscillate. Therefore the main problem has been how to explain the fact that about 70 per cent of the stars in the strip are observed to be constant to within 0.01 mag. In Section 1 we outlined the theory of Baglin (1972), who showed that the observed correlation between pulsation, rotation, and metallicism for main-sequence stars can be explained very satisfactorily as the result of a competition between diffusion and mixing processes. Stars with well-mixed outer layers are pulsationally unstable, while stars containing outer zones where the separation processes have succeeded in removing practically all helium from the main driving layers become stable and tend to develop metallicism.

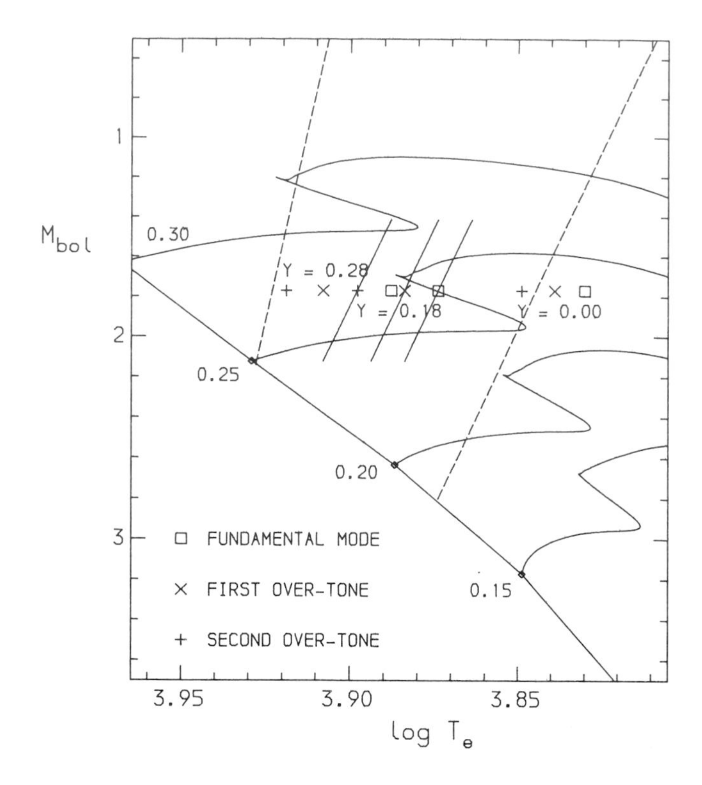

Fig.1. Theoretical blue edges for Population I (Z = 0.02) models of M = 1.5 and 2.0 $M_\odot$ as given by Cox et al. (1973). Full lines mark the edges for helium content Y = 0.18. The observed boundaries of the δ Sct instability strip are indicated (dashed lines). Full curves show ZAMS and standard evolutionary tracks for (X,Z) = (0.70,0.03) and log $M/M_\odot$ = 0.15,0.20,0.25,0.30, given by Hejlesen (1975).

Recently, Vauclair et al. (1974) have performed detailed computations of the diffusion of helium in the outermost layers of Population I models of 1.5 and 2.0 $M_\odot$. They took the changes in the structure of the models due to the helium diffusion itself into account, assuming that the two convective zones are linked by overshooting. Their result for the 1.5 $M_\odot$ model is that the helium content in the He^+ ionisation zone is reduced from Y = 0.28 to 0.10 in a time scale of 1.5×10^6 years. Since this is less than 1% of the nuclear time scale for such main-sequence stars, the diffusion processes in a real star will have proceeded much further. Thus such stars will appear stable, provided that no mixing occurs.

Although this simple theory for the correlation of pulsation, metallicity, and rotation correctly describes the properties of most main-sequence stars, a detailed discussion raises several difficult problems (Breger, 1972). The relations between rotation, mixing, and variability are discussed in more detail by A. Baglin in the next paper of the present volume.

Further it is not known if the mixing/separation hypothesis can explain the presence in the δ Scuti region above the main-sequence of relatively high amplitude variables, that all show slow rotation, together with non-variables that are often fast rotators. Clearly further theoretical studies of these complicated effects, also including rotation, are urgent. But it is well known that it is at present almost impossible to include rotation with meridional circulation in standard computations of stellar evolution. Mixing and separation processes are also of great interest in connection with two outstanding and until now unsolved problems in the standard theory of stellar evolution: The solar neutrino problem and the problem of the isochrones above the gap in the upper main-sequence in galactic clusters containing evolved stars of mass 1.5 $M_{\odot}$ or more (Maeder, 1974). Perhaps the assumption of totally unmixed evolution in radiative regions also in long time-scales is doubtful in several cases, and clearly much more detailed studies of mixing and separation processes should be included in many calculations of stellar evolution.

Compared with other types of variables typical δ Sct stars have very small amplitudes. This seems reasonable in view of the weak excitation (Chevalier, 1971; Dziembowski, 1974). However, the growth-times, although long compared to those of other Cepheid variables, are short compared to the main-sequence life-time. Non-linear investigations are needed to settle the question of the expected limiting amplitudes. But the initial value method (e.g. Christy, 1966) can not be applied successfully to δ Sct models. At present it is not known whether the small amplitudes are in agreement with the assumption of excitation of radial oscillations by the κ-mechanism. We return to these problems in more detail in the discussion of RRs stars.

Oscillations in non-radial modes would probably always result in small amplitudes, and are in this respect an attractive explanation for the group of low amplitude variables. Dziembowski (1974) showed that main-sequence stars that are unstable against radial modes are also unstable against many non-radial modes with similar growth rates. Dziembowski suggested simultaneous excitation of many non-radial modes in δ Sct stars.

Other explanations of the observed complicated light- and velocity-curves have also been offered. Fitch (1970) suggested that the complex variations observed in individual cases are due to one or more of the mechanisms: (a) tidal modulation of the basic pulsation by a faint companion in a binary orbit, (b) modulation of the primary pulsation by magnetic coupling with a companion, (c) excitation of one or more non-radial modes by rotational coupling with the intrinsic radial pulsation, (d) composite variation due to a binary pair of intrinsic variables, (e) simultaneous excitation of two successive radial pulsation modes in the same star. Le Contel et al. (1974) suggested that non-linear atmospheric effects play a role.

How can we decide whether the basic pulsation is radial or non-radial, or which complications are present in individual cases? The first problem can presumably be handled theoretically in a few years, while the second one seems more difficult. At present comparison of several frequencies in the observed complicated light-curves with the frequency spectrum calculated theoretically for the above mentioned mechanisms, seems to be the only potentially powerful possibility. When more frequencies are known in a star, their ratios and differences can be compared with calculated ones, and sometimes a clear decision may be possible. If, for instance, two radial modes are simultaneously excited in a Population I star, the period ratios will be $\Pi_1/\Pi_0 = 0.75$ to 0.76, $\Pi_2/\Pi_1 \cong 0.80$ etc. (Petersen and Jørgensen, 1972), while non-radial modes give other well defined period ratios (Ledoux,

1974; Dziembowski, 1974; see below). Another example is the equal frequency split found in the analysis of 1 Mon by Shobbrook and Stobie (1974), which was interpreted by Warner (1974) as a non-radial pulsation involving a rotational perturbation. Today very few low amplitude variables have unambiguously determined multiple periodicities (Fitch and Szeidl, 1975), and theoretical investigations of complicated systems have just begun.

Cox (1974) has pointed out that Castor's results (Castor, 1970) of linearized calculations of radial pulsations of δ Sct models may pose a difficulty. Castor found that e-folding times for amplitude changes are much smaller for the third and fourth over-tone than for the low order modes, and similar results have been obtained by Dziembowski (1974) and Stellingwerf (1975c). This indicates (but does not prove - non-linear calculations are necessary to solve the problem) that δ Sct stars according to pulsation theory should oscillate with the corresponding small periods. However, the observations seem to agree best with fundamental or low order over-tone pulsations. Also the e-folding times of non-radial p-modes for two 1.75 $M_{\odot}$ main-sequence models given by Dziembowski (1974) show this unexpected dependence on the order of the p-modes.

If δ Sct stars oscillate in such high order modes their frequency spectrum is expected to be complicated. Table 2 gives theoretically calculated period ratios for radial modes alone, non-radial modes alone, and for simultaneously excited radial and non-radial p-modes. The period ratios given for radial modes are valid to an accuracy of about 1% for all Population I variables of evolutionary masses in the lower Cepheid instability strip. Radial periods depend little on the detailed internal structure of the model, because the oscillation amplitudes are very small below relative radius about 0.5. Only very near ZAMS, where few variables are situated, an increase of 2 - 4% in the Q values for radial modes occur, due to the increasing importance of the central region (Jørgensen and Petersen, 1974). Here the ratios of over-tone to fundamental mode periods are up to 3% higher than those given in the table.

Table 2. Theoretically determined period ratios for radial oscillations and non-radial p-modes (with $l = 4$) of δ Sct models.

Radial modes		Radial modes of order				
Order	$Q \times 10^4$	0	1	2	3	4
0	333	1.000				
1	252	0.757	1.000			
2	203	0.610	0.806	1.000		
3	170	0.511	0.675	0.837	1.000	
4	146	0.438	0.579	0.719	0.859	1.000
5	128	0.384	0.508	0.631	0.752	0.877

Model	Non-radial p-modes		Non-radial p-modes of order			
	Order	$Q \times 10^4$	1	2	3	4
ZAMS	1	240	1.000			
	2	194	0.807	1.000		
	3	161	0.670	0.830	1.000	
	4	137	0.573	0.710	C.856	1.000
	5	120	0.501	0.621	0.749	0.875
X_c=0.08	1	214	1.000			
	2	196	0.916	1.000		
	3	161	0.752	0.822	1.000	
	4	138	0.644	0.704	0.857	1.000

Table 2 (continued)

Model	Non-radial p-modes		Radial modes of order and Q x 10^4					
			0	1	2	3	4	5
	Order	Q x 10^4	333	252	207	174	149	130
ZAMS	1	240	0.719	0.950	1.157	1.374	1.607	1.845
	2	194	0.581	0.767	0.934	1.109	1.297	1.490
	3	161	0.482	0.637	0.775	0.920	1.076	1.236
	4	137	0.412	0.545	0.663	0.787	0.921	1.058
	5	120	0.361	0.477	0.580	0.689	0.806	0.925
X_c=0.08	1	214	0.643	0.850	1.034	1.228	1.436	1.650
	2	196	0.589	0.778	0.947	1.124	1.315	1.510
	3	161	0.484	0.639	0.778	0.924	1.080	1.241
	4	138	0.414	0.548	0.666	0.791	0.925	1.063

The frequency spectrum for non-radial modes is much more complicated. Firstly, non-radial periods depend significantly on the detailed internal structure of the models, due to the relatively high amplitude in the central region for non-radial modes. Secondly, the value of the l parameter of the chosen spherical harmonic is important. And finally the non-radial spectrum includes besides the p-modes also the f-mode and the g-modes. Dziembowski's growth rates indicate that it is reasonable to concentrate on p-modes of order less than about 4, and the observed δ Sct periods also correspond to such oscillations. If we further restrict ourselves to fairly small l ($l \leq 4$) and to main-sequence models, rather well defined period patterns result. The Q values and period ratios of Table 2 are based upon the frequencies given by Dziembowski (1974) for two main-sequence models of 1.75 $M_{\odot}$, a zero-age model (ZAMS) and a model evolved to central hydrogen content equal to 0.08 ($X_c = 0.08$). Dziembowski (1975) reports work in progress for post main-sequence models.

3. RRs Variables

3.1 Evolutionary stage

The nature of RRs variables has been widely debated. We first review the following possibilities: RRs stars are (a) as δ Sct stars main-sequence or immediately post main-sequence stars, (b) post-red-giant stars of mass $M \cong 0.5\, M_\odot$ probably belonging to intermediate Population II, (c) stars of mass $M = 0.20 - 0.25 M_\odot$ perhaps formed in late stages of the evolution of close double star systems now contracting towards the white dwarf region, (d) blue stragglers, (e) (similar to) RR Lyrae stars of Population I; and then discuss (a), which at present seems to be most popular, in more detail.

(a) Eggen (1970) concluded that "there appears to be no obvious argument against the view that most of the ultrashort-period variables, regardless of amplitude, are stars of $M = 1.5$ to $2.5\, M_\odot$ in the shell hydrogen-burning stage of their evolution. Their presence as late evolvers in the old disk population, like that of other blue stragglers, still needs explanation, but it seems unlikely that they represent advanced evolution of lower mass stars". Chevalier (1972) also concluded that there is no compelling need to suppose that RRs variables differ basically from δ Sct stars, emphasizing the weakness of the arguments for (1) large space motions corresponding to intermediate Population II, (2) masses about $M = 0.5\, M_\odot$ or smaller, and (3) a low metal content ($Z < 0.005$) inferred from period ratios (See (b) below). Recently Bessell (1974) found no clear distinction between RRs stars and δ Sct stars on the basis of measured abundance, gravity and temperature; and from similar information derived from intermediate band photometry for 9 RRs stars McNamara and Langford (1974) concluded that RRs stars are post main-sequence stars that have not yet entered the red-giant stage.

(b) Kippenhahn (1965) discussed the stability of a model of 0.4 solar mass on the main-sequence and found 100 times more excitation than in models of normal mass ($\cong 1.5\,M_\odot$). He concluded that such models could correspond to RRs variables, and that these stars must be evolved stars which have lost most of their mass. McNamara (1965) investigated the position of RRs stars in the HR diagram and suggested that they are descendants of red-giant stars of Population I, evolving from the right to the left through the instability strip. Bessell (1969b) found from space motions and abundances that RRs stars are old disk stars intermediate between young disk Population I and old disk Population II. He also derived low masses and suggested mass-loss and perhaps mixing. These views were supported by Petersen and Jørgensen (1972), who found that the observed period ratios for RRs variables indicate a considerable deficiency in heavy elements ($Z < 0.005$) relative to the sun. Percy (1975) based his linear stability analysis for RRs stars on models of mass $M = 0.5\,M_\odot$. The derived blue edges of the instability strip (for a helium content $Y = 0.28$; see below) agree well with the observed edge. Percy determined the position of the variables in the HR diagram from a $T_e - \langle B - V \rangle$ calibration and the formula

$$0.3\,M_{bol} = \log Q/\Pi - 0.5 \log M/M_\odot - 3 \log T_e + 12.70 \qquad (3)$$

using the observed periods, the theoretically well determined pulsation constants, and the assumed mass $M = 0.5\,M_\odot$. Comparing Percy's results with the observed red boundary of the instability strip as given by Breger (1972), it is seen that about half of the variables are situated to the right of the strip (by up to $\Delta \log T_e = 0.03$). This may be due to (1) that the red boundary of the instability strip has lower T_e for lower masses, (2) systematic errors in the temperature calibration, or (3) that the assumption of a low mass is wrong. It is not possible today to decide whether (1) or (2) or (3) is correct. (1) can not be investigated successfully, because it is not yet possible to take convection into account properly in pulsation analysis. An uncertainty $\Delta \log T_e = 0.03$ is not much more than expected, but we note that an assumed mass of $2\,M_\odot$ gives a higher luminosity

($\Delta M_{bol} = -1$), bringing the variables inside the instability strip defined by δ Sct stars.

(c) Comparing results of linear non-adiabatic pulsation analysis of full stellar models of mass $M = 0.20 - 0.25\ M_\odot$ with observational data for RRs variables, Dziembowski and Kozlowski (1974) concluded that such very low mass models are plausible for RRs stars. The models have a degenerate helium core and a hydrogen-burning shell, and represent pre-white-dwarf evolution. Such objects may originate by mass exchange in close binary systems or possibly by mass loss from single stars. Since it seems rather unlikely that all RRs variables are members of close binaries, Dziembowski and Kozlowski prefer the last possibility. The blue edges for these very low masses agree well with those derived by Percy for $M = 0.5\ M_\odot$ and those based on evolutionary masses (see below). But the position in the HR diagram based on $M = 0.2\ M_\odot$ gives rise to the same problems as those just described for models of $M = 0.5\ M_\odot$. The approach of Dziembowski and Kozlowski seems to be the only alternative to the simple possibility (a) which can produce RRs models that are not in obvious conflict with the presently accepted theory of stellar evolution.

(d) Eggen (1970, 1971) suggested that some ultrashort-period Cepheids (RRs) are blue stragglers of the old disk population, while SX Phe, having extreme space motion and metal deficiency, could be similar to the blue stragglers of the halo population. The evolutionary status of the blue stragglers is not known. Such stars may be formed in binary systems with mass exchange (McCrea, 1964; Refsdal and Weigert, 1969; Refsdal et al., 1974) or by mixing (Rood, 1970; Petersen, 1972).

(e) RRs stars are separated from the c-type RR Lyrae variables by their (1) lower periods, (2) smaller amplitudes, and (3) often more irregular light-curves. With very few exceptions all Cepheids having $\Pi < 0.25$ days are found to be RRs stars. Although the RRc and RRs groups are well separated, it seems that the RRs stars of $\Pi = 0.20 - 0.24$ days have luminosities $M_{bol} \cong 1$ and also other properties very similar to the RR Lyrae group. It is therefore natural to suspect a close relationship between these groups. Kippenhahn (1965) noted that RRs stars may be some kind of RR

Lyrae stars for Population I, and Breger (1969) suggested that RRs stars correspond to the RR Lyrae stars in the sense that they are similar evolved low-mass stars. McNamara (1974) suspected the strong line RR Lyrae stars to be an extension of the RRs group.

From the point of view of stellar evolution theory the interesting point here is whether (some) RRs stars, as the RR Lyrae group, are in an evolutionary phase following the red-giant phase and the helium flash, or not. Such RRs stars would be basically different from δ Sct stars. However, present knowledge of helium-burning stellar models seems to require that all helium-burning stars must have a helium core of mass $M_c \gtrsim 0.5\,M_\odot$ and an absolute magnitude $M_{bol} \lesssim 1$. RRs stars with $M_{bol} = 1-4$ probably can not have the same structure as RR Lyrae stars. We conclude that the similarity in properties of RRs and RRc variables seems to be fortuitous from this point of view; stellar evolution theory does not predict a continuation of the RRc group to lower luminosities and smaller periods. But the similarity can probably be understood from pulsation theory.

Is the simple assumption that RRs stars are main-sequence or early post main-sequence stars evolving according to standard evolution theory in agreement with present observational data? It is well known that SX Phe presents a problem, because the luminosity derived from the reasonably reliable trigonometric parallax indicates a position in the HR diagram far below the ZAMS. The effective temperature of SX Phe has been given $\log T_e = 3.876 \pm 6$ by Bessell (1969b), $\log T_e = 3.895 \pm 5$ by Jones (1973) using a temperature calibration of his β_1 index, and $\log T_e = 3.880$ by Percy (1975). We adopt $\log T_e = 3.88 \pm 0.01$, and following the discussion of Fitch (1970) we take $M_{bol} = 4.0 \left\{ {+0.7 \atop -0.9} \right.$. In Fig.2 the error box for SX Phe is compared with results of stellar evolution calculations for hydrogen content $X = 0.70$, by Hejlesen (1975). It is seen that the masses (and ages) possible for SX Phe depend very much on the content of heavy elements, Z, assumed for the star. A standard Population I composition is not allowed, while a solar composition, $Z_\odot = 0.017$, and more metal deficient compositions are possible. Table 3 gives the mass and

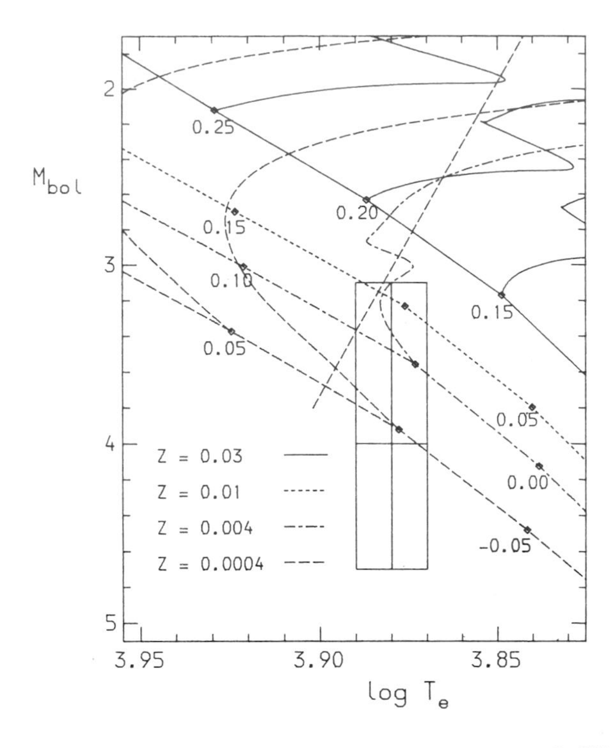

Fig.2. Comparison of the observed position of SX Phe in the HR diagram (see text) with results of stellar evolution calculations by Hejlesen (1975) for hydrogen content X = 0.70 and a large interval in content of heavier elements Z. Zero-age main-sequences are given for 4 Z values; numbers attached to the curves give $\log M/M_\odot$. A few evolutionary tracks and the cool boundary of the instability strip extended towards smaller luminosities are also shown.

Table 3. Mass and age for SX Phe

Z	Deficiency factor $Z_\odot/Z$	Mass (unit : $M_\odot$)	Maximum age (unit: 10^9y)
0.03	0.6	-	-
0.01	1.7	1.2	1.5
0.004	4	1.1	3
0.0004	40	0.9-1.0	7

the maximum age for models in the uncertainty box for a large Z interval. We note that only models of $X = 0.70$ have been considered here, but there is no evidence for large deviations from this hydrogen content among main-sequence stars.

The observed value of Preston's ΔS indicates a very small Z, and Bessell (1969b) found the Fe abundance of SX Phe to be less than in η Lep by at least a factor of 4. Assuming η Lep to have solar abundance this gives a deficiency factor of more than 4, which agrees well with the discussion of Jones (1973). Recently Jørgensen (1975) found a metal deficiency relative to the Hyades about a factor of 10 from uvbyβ photometry. Since this agrees with Bessell's result, we trust in the following a deficiency factor 4 relative to the sun, i.e. $Z = 0.004$ for SX Phe. Table 3 then provides a mass $M = 1.1\ M_\odot$ and a maximum age 3×10^9 years. We note that the absolute magnitude for SX Phe given by Percy (1975), when corrected to $M = 1.1\ M_\odot$, becomes $M_{bol} = 2.8$, which corresponds to a position 0.3 mag above the upper boundary of the box of Fig.2 and gives an age 3×10^9 years. Eggen (1971) found $M_V = 3.1$, assuming SX Phe to be a member of Kapteyn's star group. Thus it seems possible to understand the most extreme star in the RRs group in terms of standard evolution theory.

In Table 4 the characteristics of SX Phe for the assumptions (a), (b), and (c) are given. Identifying the primary period 0.05496 days with the fundamental mode and using $Q_o = 0.032$ days, the pulsation criterion gives the mean density. Using the assumed mass, radius and gravity follow. And Eq.(3) gives M_{bol} to an accuracy better than about 0.15 mag, as $\log T_e = 3.88 \pm 0.01$.

Table 4. Three possibilities for the nature of SX Phe.

Possibility	$M/M_\odot$	$R/R_\odot$	log g	M_{bol}
(a)	1.10	1.48	4.14	2.70
(b)	0.50	1.14	4.03	3.27
(c)	0.25	0.90	3.93	3.77

An observational distinction between these possibilities requires determinations of log g to an accuracy better than 0.1 or M_{bol} better than 0.5 mag. (a) is slightly favoured by the observed period ratio (see below), while (b) is suggested from the investigation by Jørgensen (1975) in this volume, and (c) agrees well with many observational data. A similar table for AI Vel shows that only a very small mass $\cong 0.2\ M_{\odot}$ can explain the observed gravities and absolute magnitudes. Then, however, it seems impossible to identify the observed period ratio, 0.773, with Π_1/Π_o.

3.2 Excitation Mechanism. Period Ratios

RRs variables clearly oscillate in radial modes as classical Cepheids and RR Lyrae stars, but they have often complicated light-curves due to simultaneous excitation of two or more modes. We have already mentioned the extensive stability analysis for models of normal masses $M = 1.5 - 2.0\ M_{\odot}$ (Cox *et al.*, 1973), models of $0.5\ M_{\odot}$ (Percy, 1975) and models of $M = 0.20 - 0.25\ M_{\odot}$ (Dziembowski and Kozlowski, 1974). In Fig.3 the blue edges of the instability strip derived in these investigations for typical Population I compositions ($Z = 0.02 - 0.03$, $Y = 0.27 - 0.28$) are compared. The blue edges of the high mass models are situated at slightly higher temperatures (for same luminosity) as expected, and the differences in log T_e (at constant luminosity) for the fundamental and first harmonic blue edges are less than 0.02 - 0.03. This is scarcely more than the differences to be expected from the small differences in composition parameters, different interpolations in opacity tables etc.

It seems unlikely that linear investigations can explain why RRs variables have higher amplitudes than δ Sct stars. Even if, for certain combinations of model parameters, one found a strong excitation (in the sense that the pulsation growth times were relatively short, e.g. 1000 periods) one could not be sure that this also secures a large limiting amplitude for such models, although it is an indication.

Christy (1966) found that non-linear investigations by the initial value method are not feasible for models of a relatively

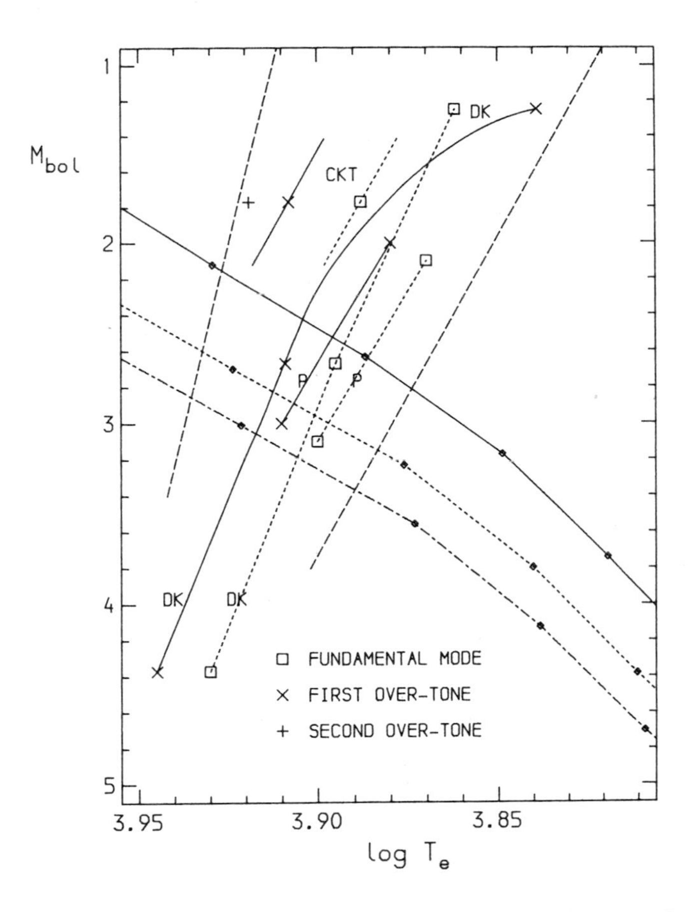

Fig.3. Blue edges of instability regions for typical Population I compositions derived by Dziembowski and Kozlowski (1974, DK) for models of 0.20 – 0.25 $M_\odot$, by Percy (1975, P) for models of 0.5 $M_\odot$, and by Cox et al. (1973, CKT) for models of 1.5 – 2.0 $M_\odot$. Fundamental mode blue edges are shown as dotted lines, and first overtone edges as full lines. For reference 3 ZAMS curves and the boundaries of the observed δ Sct instability strip are also given.

low luminosity to mass ratio, $L/M < 30\ L_\odot/M_\odot$, e.g. for δ Sct models. Such model have growth times of the order 10^6 periods; clearly excessive computer time will be used in direct simulation of the motions to limiting amplitude.

The non-linear eigenvalue approach of Baker and von Sengbusch (1969) and Stellingwerf (1974, 1975b) overcomes this difficulty, and furthermore this method provides detailed information on mode behaviour at limiting amplitude. Stellingwerf has analysed models for RR Lyrae stars and double-mode Cepheid variables of fundamental period 2 - 4 days. These stars show a complicated mode behaviour very similar to the RRs pattern. It seems likely that investigations by the non-linear eigenvalue method of models for δ Sct and RRs stars could provide valuable new information, at least for the high amplitude pulsations. In preliminary non-linear calculations for models attempted to describe AI Vel ($Z = 0.005$, $M = 0.4\,M_\odot$, $T_e = 7400\ ^\circ K$, $M_{bol} = 2.54$) and δ Sct ($Z = 0.02$, $M = 2\,M_\odot$, $T_e = 7000$, $M_{bol} = 0.875$) Stellingwerf (1975c) finds light amplitudes of at least 1.0 mag and velocity amplitudes more than 60 km/sec, which is too much. Stellingwerf's preliminary conclusion is that some source of dissipation, perhaps convective motions, is present in actual stars, but not included in the models. Stellingwerf is continuing this investigation and has encountered no problems with the numerical technique.

Period Ratios

Accurate Q values for δ Sct and RRs models determined from linear analysis have been given by several authors (Cogan, 1970; Chevalier, 1971; Petersen and Jørgensen, 1972; Dziembowski and Kozlowski, 1974; Jørgensen and Petersen, 1974). In principle, comparison with observed pulsation constants tells in which mode a variable oscillates. However, present available observational accuracy does not allow definite answers for RRs stars (Section 2.1).

Petersen and Jørgensen (1972) studied the ($\log \Pi_0$, Π_1/Π_0) diagram, and showed that this diagram provides potential powerful means to determine mass and chemical composition for double mode pulsators having two observationally accurately known periods. They concluded that the observed period ratios for RRs variables $\Pi_1/\Pi_0 = 0.768 - 0.778$ indicate that RRs stars have $Z < 0.005$ and thus cannot belong to Population I. This result was questioned by Chevalier (1972) and Baglin <u>et al.</u> (1973), who emphasized that neither the theoretical calculations nor the observed periods were accurate enough to allow distinction between period ratios near 0.77 for RRs variables and the values 0.750 - 0.760 expected for Population I models.

The weakest point in the theoretical analysis is the application of the linear, adiabatic theory for the derivation of Q values. However, in the meantime linear non-adiabatic Q values (e.g. Cox et al., 1972; Dziembowski and Kozlowski, 1974) and Q values derived in full non-linear calculations for models corresponding to double mode Cepheids (Stellingwerf, 1975a, 1975b) have been found to agree well with the linear adiabatic pulsation constants. In the present context Stellingwerf's results are especially interesting. Analysing a mixed-mode RR Lyrae model of fundamental period 0.96 days, he found linear and non-linear periods to agree to an accuracy of 0.2% for the fundamental mode and 0.5% for the first over-tone. He also showed that the standard procedure used by observers to determine periodicities in observed light curves is correct in the sense that the periods derived by this method from the calculated light curve (with beats) are precisely the same as those corresponding to pure fundamental or first harmonic oscillations of the model. Due to the non-linear character of the problem this is not a priori evident. It now seems that the small differences in Π_1/Π_0 discussed above are significant both theoretically and observationally.

Among the RRs variables SX Phe has the highest period ratio ($\Pi_1/\Pi_0 = 0.778$) and the largest metal deficiency ($Z \cong 0.004$). From the calibration curves in Fig.4 of Petersen and Jørgensen (1972) or in Fig.4 of Dziembowski and Kozlowski (1974), or from data given by Jørgensen and Petersen (1974) it is seen that these values are in agreement with a mass $M \cong 1\ M_\odot$ for SX Phe. For very small masses, $M \cong 0.2\ M_\odot$, the calibration curves indicate a considerably smaller period ratio. However, it is not known how much reasonable changes in X and Z can change the period ratio.

If the other RRs variables are slightly metal deficient and have $M = 1.5 - 2.0\ M_\odot$, we expect period ratios near the observed ones (0.768 - 0.773). But if they have $M \cong 0.2\ M_\odot$ and a nearly normal Population I composition, the identification of the period ratio with Π_1/Π_0 seems impossible, as just mentioned for AI Vel. Clearly much work remains to be done on these problems.

4. Conclusions

(1) Comparisons of observational data for δ Sct stars with theoretically calculated evolution and pulsation properties of stellar models all agree that δ Sct variables are normal young disk stars of Population I evolving in the main-sequence or in the shell hydrogen-burning post main-sequence phases.

(2) Linear stability analyses of δ Sct models (Cox *et al.*, 1973) give fundamental and first and second over-tone blue edges of the instability strip that agree very well with the observed strip in the HR diagram, when a standard helium content is assumed. Q values determined by Breger (1975a) indicate that the high temperature δ Sct stars near the blue edge oscillate in the first and second over-tone, while the cooler variables oscillate in the fundamental mode; just as expected.

(3) The presence in the instability strip of non-variable stars, often with Am characteristics, have been explained in terms of separation processes (Baglin, 1972; Vauclair *et al.*, 1974). In the main-sequence rotation seems to control the separation and mixing processes; slowly rotating stars become non-variable Am stars, while relatively fast rotators become typical δ Sct variables.

(4) Non-radial modes are also excited in δ Sct models (Dziembowski, 1974). The observed low amplitudes can be taken as an argument for the presence of non-radial modes rather than radial modes. We suggest comparison of observed multiple periodicities with the theoretically determined period patterns of radial and non-radial oscillations as a means to distinguish between these possibilities. Period patterns may also provide information on other complicating effects, e.g. coupling of pulsation and rotation.

(5) It is not clear whether RRs variables are of the same nature as δ Sct stars or quite different, probably in a very advanced evolutionary stage. We have shown that it is possible to explain most observed properties of SX Phe from the assumption that this star, as δ Sct stars, is a main-sequence or immediately post main-sequence star.

(6) The similarity of the pulsations of RRs stars to those of other types of double mode Cepheids, and their observed period ratios, show that RRs variables pulsate in radial modes. Linear stability analyses agree well with the observed position in the HR diagram.

(7) At present the reason for the difference in amplitude between typical δ Sct and RRs variables is not known. Non-linear investigations by the initial-value method are not feasible for δ Sct and RRs models, but the new powerful eigenvalue method may provide important information on limiting amplitudes in the near future.

(8) The observed period ratio of SX Phe agrees with a mass $M \cong 1\ M_{\odot}$ for this star, and a recent investigation by Stellingwerf (1975b) of double mode behaviour of an RR Lyrae model indicates that the small differences in period ratio used in this type of analysis are significant. Clearly non-linear analysis of RRs models must be available before results derived from period ratios can be trusted.

(9) We finally mention some important developments expected in the near future : (i) Non-linear, radial survey. An investigation by Stellingwerf (1975c) is in progress. (ii) Linear, non-radial survey. Dziembowski (1975) has obtained detailed results for main-sequence and post main-sequence δ Sct models. (iii) Comparison of detailed models with data for individual objects may soon provide more reliable information on the presence of radial or/and non-radial oscillations in δ Sct stars and the nature of RRs variables.

Acknowledgements. The author wants to thank Drs. M. Breger, W. Dziembowski, W.S. Fitch, and R.F. Stellingwerf for providing unpublished material, Dr. H.E. Jørgensen for many discussions during this work, and Dr. P.M. Hejlesen for help and advice in the application of the UNIVAC 1110 computer of the RECKU computing center.

References

Baglin, A. 1972, Astron. & Astrophys. 19, 45.

Baglin, A., Breger, M., Chevalier, C., Hauck, B., Le Conel, J.M., Sareyan, J.P., Valtier, J.C. 1973, Astron. & Astrophys. 23, 221.

Baker, N., von Sengbusch, K. 1969, Mitt. Astron. Gesell. 27, 162.

Bessell, M.S. 1969a, Astrophys. J. Suppl. 18, 167.

Bessell, M.S. 1969b, Astrophys. J. Suppl. 18, 195.

Bessell, M.S. 1974, IAU Symposium No.59, 63.

Breger, M. 1969, Astrophys. J. Suppl. 19, 99.

Breger, M. 1972, Astrophys. J. 176, 373.

Breger, M. 1974, Astrophys. J. 192, 75.

Breger, M. 1975a, IAU Symposium No.67, 231.

Breger, M. 1975b, to be published in Astrophys. J.

Breger, M., Kuhi, L.V. 1970, Astrophys. J. 160, 1129.

Castor, J.I. 1970, Bull. Am. Astr. Soc. 2, 302.

Chevalier, C. 1971, Astron & Astrophys. 14, 24.

Chevalier, C. 1972, Ages of δ Scuti and AI Velorum Stars, in Stellar Ages, IAU Colloquium No.17, Ed. G. Cayrel de Strobel, A.M. Delplace, Observatoire de Paris.

Christy, R.F. 1966, Ann. Rev. Astron. and Astrophys. 4, 353.

Cogan, B. 1970, Astrophys. J. 162, 139.

Cox, J.P. 1974, Rep. Prog. Phys. 37, 563.

Cox, J.P., King, D.S., Stellingwerf, R.F. 1972, Astrophys. J. 171, 93.

Cox, A.N., King, D.S., Tabor, J.E. 1973, Astrophys. J. 184, 201.

Crawford, D.L. 1970, IAU Symposium No.38, 283.

Dickens, R.J., Penny, A.J. 1971, Mon.Not.R.Astr.Soc. 153, 287.

Dziembowski, W. 1974, Communication at 19th Liège International Astrophysical Colloquium.

Dziembowski, W. 1975, private communication.

Dziembowski, W., Kozlowski, M. 1974, Acta Astronomica 24, 245.

Eggen, O.J. 1970, Publ. Astron. Soc. Pacific 82, 274.

Eggen, O.J. 1971, Publ. Astron. Soc. Pacific 83, 762.

Fitch, W.S. 1970, Astrophys. J. 161, 669.

Fitch, W.S., Szeidl, B. 1975, to be published.

Gupta, S.K. 1973, Observatory 93, 192.

Hejlesen, P.M. 1975, to be published.

Jones, D.H.P. 1973, Astrophys. J. Suppl. 25, 487.

Jørgensen, H.E. 1975, to be published.

Jørgensen, H.E., Petersen, J.O. 1974, Astron. & Astrophys. 35,215.

Kippenhahn, R. 1965, IAU 3rd Colloquium on Variable Stars, Bamberg, 7.

Le Contel, J.M., Valtier, J.C., Sareyan, J.P., Baglin, A., Zribi, G. 1974, Astron. & Astrophys. Suppl. 15, 115.

Ledoux, P. 1974, IAU Symposium No.59, 135.

Maeder, A. 1974, Astron. & Astrophys. 32, 177.

McNamara, D.H. 1965, IAU 3rd Colloquium on Variable Stars, Bamberg, 111.

McNamara, D.H. 1974, IAU Symposium No.59, 65.

McNamara, D.H., Langford, W.R. 1974, IAU Symposium No.59, 65.

Pamjatnikh, A.A. 1974, Informatsij Nautshnij 29, 108.

Percy, J.R. 1975, Mon. Not. R. Astr. Soc. 170, 155.

Petersen, J.O. 1972, Astron. & Astrophys. 19, 197.

Petersen, J.O., Jørgensen, H.E. 1972, Astron. & Astrophys. 17,367.

Refsdal, S., Weigert, A. 1969, Astron. & Astrophys. 1, 167.

Refsdal, S., Roth, M.L., Weigert, A. 1974, Astron. & Astrophys. 36, 113.

Rood, R.T. 1970, Astrophys. J. 162, 939.

Shobbrook, R.R., Stobie, R.S. 1974, IAU Symposium No.59, 67.

Stellingwerf, R.F. 1974, Astrophys. J. 192, 139.

Stellingwerf, R.F. 1975a, Astrophys. J. 195, 441.

Stellingwerf, R.F. 1975b, preprint.

Stellingwerf, R.F. 1975c, private communication.

Strömgren, B. 1966, Ann. Rev. Astron. Astrophys. 4, 433.

Warner, B. 1974, IAU Symposium No.59, 68.

Vauclair, G., Vauclair, S., Pamjatnikh, A. 1974, Astron. & Astrophys. 31, 63.

Discussion to the paper of PETERSEN

BAGLIN: Before asking the question concerning the different evolutionary stage of the two classes of variables, δ Scuti and RRs, one should define clearly their properties. The first classification in the Catalog of Variable Stars, made from a rough look at the light curve, cannot be retained. As far as the amplitudes are concerned, there seems to be a continuous transition from low to high amplitudes, rather than a clear break.

PETERSEN: I have used the definition based on amplitudes (limit: $\Delta V = 0.30$ mag). There is a continuous transition from relatively high amplitude δ Scuti stars with amplitudes near 0.30 mag to RRs stars of slightly higher amplitudes. But I mostly refer to typical low amplitude δ Scuti stars and RRs variables.

Multiple Periodic Variable Stars
IAU Colloquium No.29, Budapest, 1975

ROTATION MIXING AND VARIABILITY IN A STARS

A. Baglin

Observatoire de Nice, Le Mont-Gros, 06300 Nice

The purpose of this paper is to contribute to answer the question : why among A stars are there variable and "constant" stars ? To do this we will first review the observational properties of these stars, then discuss the theoretical models based on the interplay between microscopic diffusion and rotation, and finally discuss the "unsolved" questions !

A Review of the properties of the A stars in the instability strip (i.e. on the main sequence from A2 to F0, and somewhat later types in giants).

1 Dwarf stars

We will first restrict the problem to the main sequence stars because there the situation is clearer and more quantitative work has been done on it i.e. there exists photometric calibration which permits to compute M_{bol}, $\log T_{eff}$, ...

1.1 On the main sequence among A stars the spectroscopists have recognized long ago the "normal stars" and the "metallic line stars" (1940). I do not want to discuss in detail the definition of the Am phenomenon.

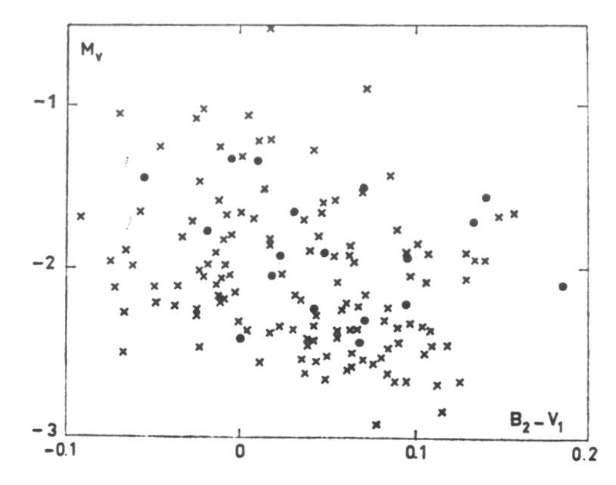

Figure 1.- Observational Hertzsprung-Russell diagram of Am and δ Scuti stars based on the photometric system of the Geneva Observatory. ⋆ Am . δ Scuti from Baglin (1972).

Following Conti (1970) "the Am phenomenon is present in stars that have an apparent surface underabundance of Ca (and/or Sc) and/or apparent overabundance of the heavier (iron group) elements". The nature of the anomalies does not vary significantly with the temperature (Smith, 1973).

From extensive search of variability in the instability strip Breger (1970) has suggested that typical Am stars are not variable, and vice et versa. This result has been confirmed up to now.

Several doubtful cases have appeared, they have generally been solved in agreement with the preceding statement. Up to now, over more than 30 Am stars studied only three cases remain doubtful to my knowledge.

HR 5491 a typical Am star has been observed variable by Bessel and Eggen (1972). However the light curve seems to be very irregular. It has been reobserved since by several authors (Breger et al, 1972; Stobie and Eggen, 1973) and found constant.

32 Vir which has been classified sometimes as an Am (Cameron, 1967) but which also is considered as magnetic by Babcock (1958) has been shown to be variable by Bartolini et al (1972). However it is

a very peculiar object; its spectrum is highly variable and the K line does not show the velocity variation of its orbital motion as the metallic lines do. It is probably a quite close binary composed of a normal A star and of an Am star (Breger, private communication).

ν^2 Dra has been proposed as a variable star by Mendoza and Gonzales (1974). However this variability needs to be checked.

1.2 The variable stars have a "normal spectrum" and normal solar abundances. With the present observational threshold (one thousandth of a magnitude) 34% of the "normal" stars have been detected as variable. I recall here Breger's result that the distribution of the amplitudes is consistent with the fact that all the normal stars are variable most of them having an undetectable amplitude.

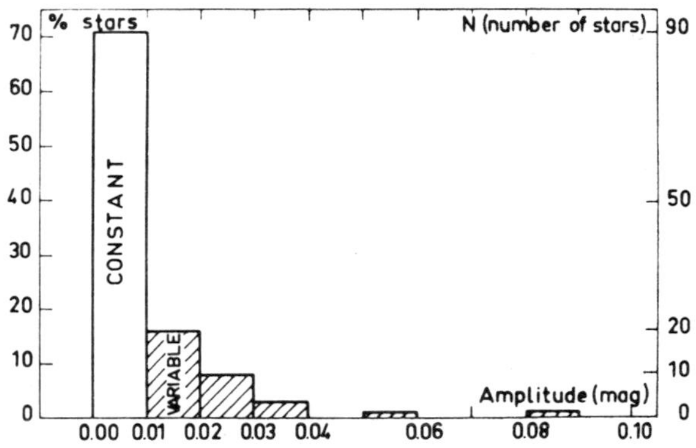

Figure 2.- Histogram of the amplitudes of known variables.
Undashed area : stars for which no variability can be established which lie in the instability strip.
From Baglin et al, 1973.

1.3 The main difference between "normal" and Am stars consists in the distribution of the rotationnal velocities. The first result of Slettebak (1955) has been confirmed by a more precise and extended work by Abt and Moyd (1973) :

$$\langle V \sin i \rangle_{A_m} = 33 \text{ km/s}$$

$$\langle V \sin i \rangle_{A_{nor}} = 141 \text{ km/s}$$

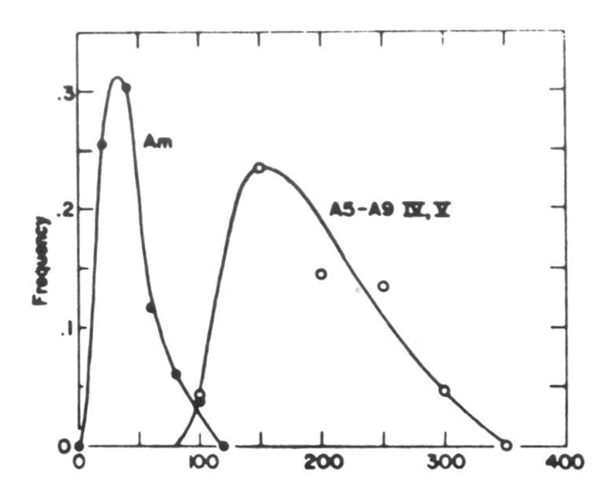

Figure 3 .- Distribution of the apparent rotational velocities in Am and normal A stars from Abt and Moyd (1973).

Smith (1971) has noted that statistically the degree of metallicism decreases as V_{Rot} increases.

As this question is of great importance for our purpose, let us comment on several difficulties.

i- There are normal stars which rotate slowly. Out of 34 variables from the catalogue of δ Scuti stars by Baglin et al (1973) , 3 have V sin i smaller than 50 km s^{-1} which is compatible with a random distribution of the inclination angles.

However HR 8584 which is a spectroscopic binary of period 2.3 days can hardly have a large inclination, its rotational velocity (32 km s^{-1}) is of the order of the synchronism velocity (25 km s^{-1}) (14 Aur, which is slightly evolved presents the same situation).

In the Pleiades the distribution of the rotational velocities is analogous to that of the field stars. However no Am has been detected. So the intrinsically slowly rotating A stars of the cluster are probably not Am.

ii- Some Am stars seem to rotate rapidly.

In the list of Abt and Moyd the fastest is HR 4646 with 95 km s^{-1}. Smith (1972) has analysed HR 7774 and showed that the most probable explanation of this system is : two rapidly rotating Am stars ($V \simeq 120$ km s^{-1}) identical in a binary of period 61 days. However this very curious system needs more attention.

Note : As all the "normal" stars are variable, there can be some difficulty in defining the rotational velocity as measured from the "half width" (see Valtier, this conference).

1.4 Another parameter which is usually called upon to distinguish Am and "normal" stars is that Am stars are members of binaries.

The periods are extremely variable from 1 to more than 100 days, i.e. the binary systems are sometimes very close and sometimes so detached that no interaction exists between them.

However as shown by Zahn (1972) the binary motion through the dynamical tide produces a braking of the rotation of stars with a convective zone . The final state is that axes of intrinsic and binary motions are parallel and rotations are synchronized.

The time scale of this process for an A star is smaller than the evolutionary time scale if the period is less than 6 days. However, observations seem to show that this critical period is longer ...

In the following we will consider that the effect of the binary nature of A stars is to produce slow rotation.

2 Subgiants and giants

2.1 With respect to the "abundance" problem the situation is less clear than on the main sequence. The extension of the Am characteristics to this domain is under investigation. The Ca anomaly seems to decrease when the gravity decreases, though the Fe underabundance does not vary and Zr/Fe increases. In addition there seem to be a large variety of stars "with abnormal spectra" called by Breger the "cosmic garbage box" in which the spectroscopists have tried to distinguish classes : δ Del, F III p, λ Boo, ...

As seen from the discrepant spectral type determinations from different authors, the visual classification is hazardous. Probably there are other mechanisms than the variation of the abundances which alter the aspect of the spectrum i.e. luminosity, variability.

Now several detailed abundance analyses are available. From the most recent works (Kurtz cited by Breger, 1974b, and Ishikawa, 1973) the following results appear :

- the shape of the H and Ca lines are generally normal for their T_{eff} and g,
- the iron abundance is normal,
- the rare earth are enhanced by a factor ∿ 3.

2 No extended search for variability has been made on these stars so that no statistical results are available.

- In the "cosmic garbage box" 9 out of 13 studied are known pulsators, generally with large amplitudes.

TABLE 2

Interesting Delta Scuti Variables With Unusual Spectra

Metallicity Index (uvby)	Metal Abundance	Spectral Classifications		
		Cowley +Jaschek	Abt +Bidelman	Morgan +Abt
(high)	All up a bit	-	-	F3IIIp
(normal)	All down a bit	δ Del	δ Del	F0IVp
(high)	-	F2III	-	F3IIIp
(high)	All up a bit	-	-	F5IIp
(high)	All up a bit	-	-	F3III
normal	-	δ Del	-	-
(normal)	-	δ Del	Poss. Am	A9IV
normal	-	A8V and A5m:	Poss. Am	A9IV?

Figure 4 .- Interesting δ Scuti stars with abnormal spectra from Breger (1974b).

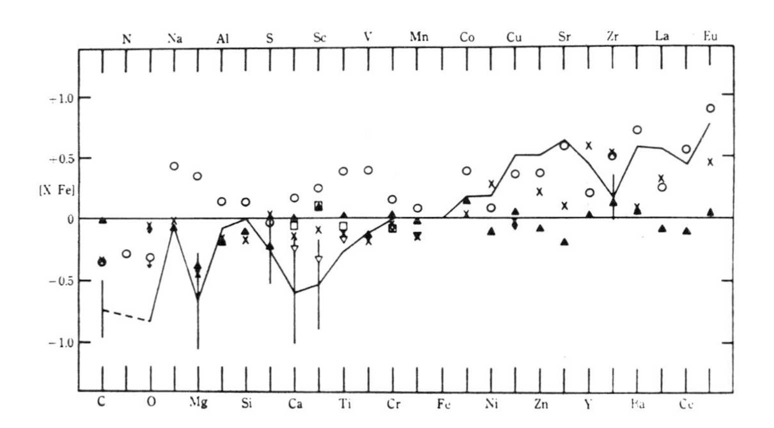

Figure 5 .- Individual abundances of δ Scuti stars from Ishikawa (1973).

- There are also pulsators with normal spectra, i.e. 14 Aur, HR 432.

2.3 Concerning the rotation parameter no definite statement can be made.

Among normal stars the pulsators can be fast or slow rotators : 14 Aur cannot have a very small sin i and V_R sin i = 30 km/s, HR 432 has broad lines corresponding to V_R sin i ≃ 110 km s^{-1}.

In the "cosmic garbage box" there seems to be a tendency to have essentially a small rotational velocity and a large amplitude.

However, if the anticorrelation between rotational velocity and amplitude close to the main sequence is still valid, the fast rotators should have small amplitude. As they are evolved, the period is longer (ex HR 432). These two conditions make them difficult to detect -

B Theoretical interpretation.

I recall the main result that the instability strip corresponds to sets of T_{eff}, g of the atmospheres for which the helium ionization zones are properly situated to trigger a vibrational instability of the whole star. The contribution of the hydrogen ionization zone is not yet well understood because it is located in the high outer layers where the motion is non linear. However, it seems that it is never the principal agent of instability (as proven by the good agreement between the observational and theoretical borders of the instability strip, which are calculated in the δ Scuti case at least, in the linear approximation).

This vibrational instability has been established for the normal cosmic helium abundance $y \simeq 0.3$ by weight.

The idea comes to attribute the disappearance of the instability (in the same ($\log T_{eff}$, $\log g$) region) to the disappearance of helium in these outerlayers, or at least to an important decrease of the He abundances. The influence of the helium content on the nature of the pulsation in the instability strip has been studied by Cox, King and Tabor (1973). The major result for the main sequence neighbourhood is that when y (in weight) becomes less than 0.2 the fundamental remains stable.

Note that the He abundance of Am stars is not known. For the hottest ones where some lines are visible, Smith (1974) has found that helium was deficient by a factor 3 to 5 in α Gem, θ Leo and Sirius. This last determination contradicts the previous analysis of Kohl (1964) who found He normal in Sirius.

Then it becomes tempting to attribute to the same process the He and metals anomalies in Am stars. The microscopic diffusion as we will see can do the job.

The last question is then : why diffusion processes can occur in slow rotator stars and not in fast ones? We will see how fast rotation can give a meansof mixing the outerlayers of these stars.

Let us now go into the details of this model and then rediscuss the observational facts.

1 Microscopic diffusion

Let us recall the general equation of diffusion in a binary gas mixture as given by Chapman and Cowling (1960)

$$\omega_1-\omega_2= -D_{12}\left[\frac{1}{c_1}\,\frac{1}{c_2}\frac{\partial c_1}{\partial r}+\frac{m_2-m_1}{c_1m_1+c_2m_2}\,\frac{1}{P}\,\frac{\partial P}{\partial r}-\frac{m_1m_2(F_1-F_2)}{(c_1m_1+c_2m_2)kT}+\alpha_{12}\frac{1}{T}\,\frac{\partial T}{\partial r}\right] \quad (1)$$

ω_i, m_i, c_i, F_i, velocity in the F direction, mass, concentration, external forces on particles of species i $(i = 1, 2)$

α_{12} coefficient of thermal diffusion

D_{12} coefficient of diffusion

In the case of a test particle (of very small abundance) in completely ionized hydrogen Aller and Chapman (1960) computed the electric field and obtained the velocity of diffusion of an ionized test-particle assuming that the only forces acting are the electric and gravitational forces. This treatement is justified for A stars as the outerlayers are almost completely devoid of. neutral atoms.

$$\omega_2 = D_{12} \left[-\frac{1}{c_2} \frac{\partial c_2}{\partial r} + k_1 \frac{1}{P} \frac{\partial P}{\partial r} + k_2 \frac{1}{T} \frac{\partial T}{\partial r} \right] \quad (2)$$

$k_1 = 2A - 2 - 1$

$k_2 = 2.54\, z^2 + 0.805\, (A - 2)$

$D_{12} = 6.62\ 10^9\ 5^{5/2}\ (n_1 M_2\ z^2 \alpha)^{-1}$

n_1 is the number density of particles.

$$M_2 = \frac{m_2}{m_1 + m_2}$$

$$\alpha = \log_e \left(1 + \left(\frac{4 d_D kT}{Ze^2}\right)^2\right) \simeq \log_e \left(\frac{4 d_D kT}{Ze^2}\right)^2 \text{ slowly varying with T}$$

$$d_D = \left(\frac{kT}{4\pi \mu_e e^2}\right)^{1/2} \quad \text{so} \quad \alpha \simeq \log_e \left[6.4\ 10^{-16} \frac{T^3}{Z^2 p_A}\right].$$

In the outerlayers of A stars where $\frac{d \log T}{d \log P} \simeq 0.5$ the ratio of the pressure term to the thermal one is

$$\simeq 0.5 \frac{k_2}{k_1} \simeq \frac{Z^2}{A}$$

i.e. of the order of 1 for helium and smaller for heavier elements.

So generally the pressure term dominates. In this case the diffusion velocity can be estimated by

$$\omega_2 \quad D_{12}\, k_1 \frac{1}{P} \frac{\partial P}{\partial r} = 10^{-22} \xi A T^{3/2} p^{-1} g \quad (3)$$

ξ is numerical coefficient close to 1 depending slightly on the composition (A, 2, μ)

In the case of helium the direct use of these formulae raises some difficulties because helium is not exactly a test particle.

The error due to the neglect of the collisions between test particles can be estimated using the second approximation of D_{12} from Chapman and Cowling (1960). For the Cosmic helium abundance $\frac{n_2}{n_1} \simeq 0.1$ it is of the order of several percents. However, the treatment of the electric field has been modified taking into account the electrons coming from ionization of helium. Work is in progress on that point by Pamjatnickh and Michaud and Montmerle (private communication).

2 Radiation pressure effect

The electric force is not always the only external force acting on the particles. Michaud (1970) introduced the idea that the radiation field coming from the interior of the star will push the different atoms upward. The strength of this effect will depend on the interaction between the radiation field and the species i.e. on the spectrum of the radiation, on the atomic structure of the species and probably on the abundance.

The upward acceleration on atoms of species 2 is

$$g_R = \frac{1}{m_2} \sum_{i,n} \frac{N_{i,n}}{N_e} \int_\nu \sigma_{i,n}(\nu) \frac{\phi(\nu)d\nu}{c} \qquad (4)$$

where i and n are the ionization and excitation states $\frac{1}{c}\phi(\nu)d\nu$ is the energy density and σ_i, n the absorption cross section of the atom in the state considered.

In the case we are interested in i.e. outerlayers of A stars, for most species the only important contribution of g_R for heavy elements comes from the lines.

Then $m_2\ g_R$ has to be considered as an external force in the expression of ω_2, and after some modifications, using the perfect gas law approximation this term is added to the pressure term changing k_1 to

$$k_R = 2A\ (1 - \frac{g_R}{g}) - 2 - 1 \qquad (5)$$

assuming that no radiation pressure acts on protons.

On this expression one sees clearly that if in a layer g_R becomes larger than g the velocity of diffusion changes sign.

3 Diffusion in A stars

Using equation (3) one can estimate the diffusion velocity when knowing T and ρ. In A stars it is of the order of 10^{-5} cm s^{-1} and the corresponding time scale $t_D = \frac{H_P}{\omega_D}$ is approximately 10^5 to 10^6 years.*

Diffusion to take place needs a stable medium at least over this period. Convection which is a very powerful agent of mixing destroys immediately the effect of diffusion. Then, diffusion can take place only in radiative zone. But acting at the outerboundary of a convective region it is able to sort

---

* This time scale corresponds to the sorting of the whole outerlayer down to the bottom of the convective zone, as its dimension is comparable to H_P.

a whole convective zone (which remains homogeneous) above.

In A stars the radiative zone intermediate between the two convective zones (see fig 6) is not stable and probably completely mixed by overshooting from the two convective zones, as proposed by Toomre et al (1975). In the homogeneous model it can take place only at the bottom of the convective zone.

But Vauclair, Vauclair and Pamjatnikh (1973) computed the influence of the diffusion of helium at the bottom of the convective zone on the structure of the models. They found that the outerlayers are deprived from helium in 10^6 years and that the second convective zone disappears (fig 6 et 7).

Then after 10^6 years, a stable star will be left with only a superficial convective zone. After this time has elapsed the diffusion processes will take place at the bottom of this zone.

Let us now examine the problems of metals for which under/over abundances have to be explained. This cannot be done with ordinary diffusion which will always have a tendency to deplete the surface of heavy elements. Then the influence of the radiation field becomes fundamental.

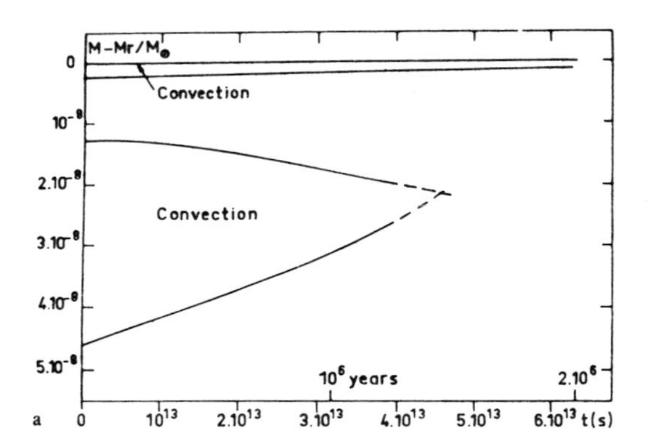

Figure 6 .- Evolution of the second helium convective zone, from Vauclair Vauclair, Pamjatnikh (1973).

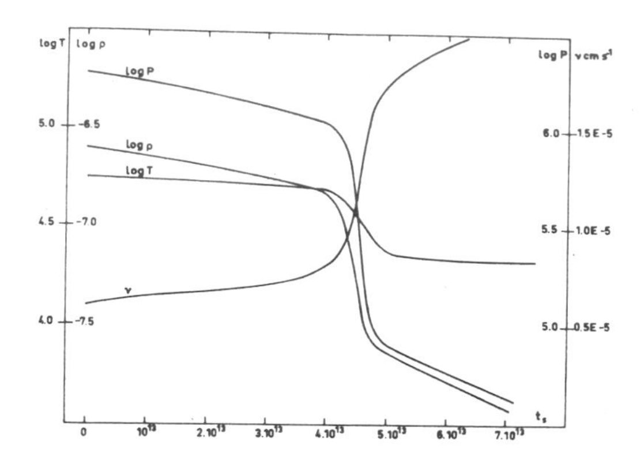

Figure 7 .- Variation of T, ρ, $V_{diffusion}$ at the lower boundary of the second helium convective zone during evolution from Vauclair, Vauclair, Pamjatnickh (1973).

At the bottom of the second ionization zone, the effect is small, the diffusion velocities are too low. For radiation pressure to be efficient the sorting has to take place at lower temperature. This is possible if the separation of metals takes place after the disappearance of the second convective zone, at the bottom of the first one.

Detailed computations of g_R are now available from the work of Kobayashi and Osaki (1973) for several elements Sc, Sr, Y, Zr (fig 8). In these computations the gravity is 10^4 and the radiation pressure force is given as a function of depth, and of the abundance.

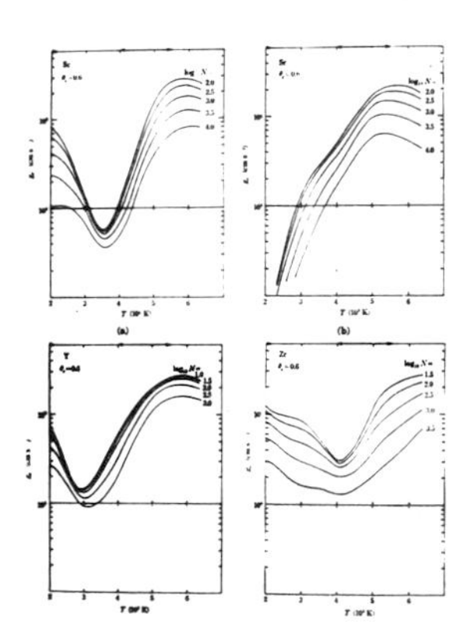

Figure 8 .- Upward acceleration due to the radiative force on Sc, Sr, Y, Zr as a function of the depth of the envelope for the case Θe = 0.6. Arrows indicate the location of the convective zones.

The authors found that "around 3.5 10^4 d°K" the radiation force acting on Sc is smaller than the gravitational force, while the radiation pressure forces acting on Sr, Y and Zr are larger than the gravitational force. Fig 9 illustrates the extreme sensitivity of the radiation force to the temperature; in particular the existence of a low g_R domain imposes drastic conditions on the temperature at the level where diffusion takes place.

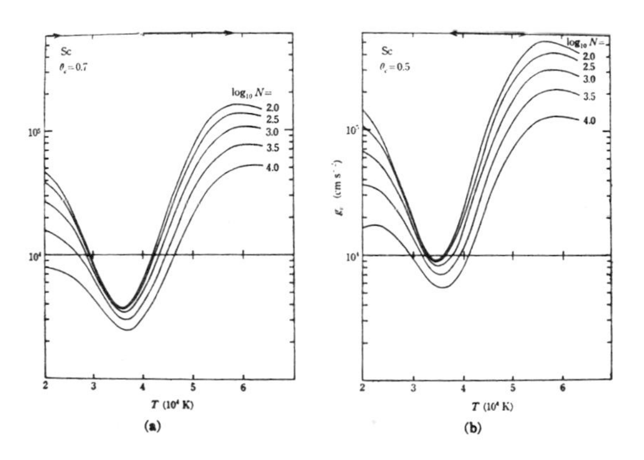

Figure 9 .- Same thing for Sc and different values of Te.

The case of calcium has not been treated in detail but due to its electronic structure it has a similar behaviour to Sc.

These results should be taken only qualitatively : the treatment needs still many improvements as for example the influence of the different ions of the same species present at the same time. In that case for which the coefficient of diffusion being proportionnal to Z^{-2} , Schatzman and Alecian (private communication) have shown that the diffusion velocity is the weighted sum of the diffusion velocities of the different ions

$$W_D = \sum_i x(i)\, w(i) \qquad (6)$$

where x(i) is the degree of ionization of the ion i, and w(i) its diffusion velocity as computed from eq (2).

The underabundance of Sc (or Ca) is the most difficult to obtain and necessitates very drastic conditions on Te (and/or g) whereas the overabundances of heavy metals seem to be more easy to explain. The discussion of the abundances of the giants showed that, though the Ca deficiency seems to disappear, the heavy metals overabundance is very frequent, even in pulsating stars.

Time scale of sorting : Velocities of diffusion are 5 10^{-5} cm s^{-1} for Ca and Sc and 10^{-4} cm s^{-1} for Sr, Y, Zr, leading to time scales of the order of 10^5 to 10^6 years.

The age of Am stars seems to be highly variable. Smith (1972) searching for Am stars in clusters found some of them in very young clusters like Ori Ic which is 10^6 years old. Then this diffusion time scale is short enough to explain all the Am.

Predicted values of over(under) abundances : As there is enough time for diffusion to proceed, the equilibrium abundances should be reached. They satisfy $\omega_2 = 0$ i.e.

$$\frac{1}{c}\frac{\partial c}{\partial r} - A\left(2-\frac{g_R}{g}\right)\frac{1}{P}\frac{\partial P}{\partial r}=0 \quad . \tag{7}$$

As A is large for metals (≃ 100) the concentration gradient will be steep and the equilibrium abundance quite extreme.

This effect can be somehow reduced by the saturation of g_R which can be seen in fig 9.

When the abundance of the element is small the acceleration is independent of it. But when the element becomes a major contributor to the opacity, the process saturates and g_R is proportionnal to N^{-1}. Eventhough, the equilibrium abundances would be much more extreme than the observed ones (30 at most). Then as discussed in Schatzman (1969) a diffusion barrier is needed.

To sum up let us summarize what the theory of diffusion is probably able to explain in A stars.

Assuming that for some reason, (see sect. III) the outerlayers of Am stars are stable it predicts that :

1) the Am envelopes are deprived from helium.
This point explains why the Am do not pulsate. Vauclair, Vauclair and Pamjatnick (1973) have shown that after 10^6 years the second convective zone, responsible for the driving of the pulsation disappears. So, after this time has elapsed the pulsation cannot be fed anymore.

2) the sign of the abundances anomalies for the most important metals are explained, though some difficulties remain in the quantitative results.

3 Mixing

Let us examine now the question : why are the homogeneous and sorted stars in the same log T_{eff}, log g domain which is the theoretical equivalent of "why are there normal and Am stars in the same region of the HR diagram"

To have variable normal stars the diffusion of helium below the second convective zone has to be prevented by some "mixing agent", which should not be present in the slowly rotating Am stars.

The main difficulty in this kind of problem is that we do not know a general criterion for stability. What has been done up to now is to list some instabilities and study conditions under which they take place. We do not know either how to describe the state of motion produced by these instabilities. So we are left with very rough estimates quite unsatisfactory.

3.1 Meridional circulation

The well known Eddington-Sweet meridional circulation present in all rotating stars is certainly in itself a way of mixing. It is due to the thermal unbalance between pole and equator. The horizontal component drives away the elements from the surface to the interior and the vertical component produces a flux of matter through the convective zone. Only in the case of solid body rotation the velocity field (Mestel, 1965) can be computed easily; the boundary of the convective zone is a singularity for the velocity field. However the overshooting (over a scale height of the order of H_p) erases this singularity.

As remarked by Baglin (1972) the horizontal flow does not stop diffusion. But the time scale is lengthened. The lengthening depends on the exact topology of the motion in the convective zone like the shape of the cells, etc., ...

The time scale of the mixing of the outer zones can be estimated by the time needed to refill with new matter an outer layer of mass ΔM. Using the value of the vertical component as given by Mestel (1965)

$$V_r = \frac{\bar{\rho}}{\rho} \frac{R}{t_{KH}} \frac{\Omega^2 R}{g} \tag{8}$$

where $\Omega, R, g, \rho, \bar{\rho}$ have their usual meaning and t_{KH} is the Kelvin Helmholtz time scale , this time is

$$t_{MC} = \frac{\Delta M}{M} t_{KH} \frac{\Omega^2 R}{g}^{-1} . \tag{9}$$

This time scale is somewhat longer than the diffusion time. For this process to work, the meridional velocity field has to be completely stable and this does not conserve the solid body rotation.

3.2 Turbulent mixing from the shear of the meridional circulation

Below the surface, there is a shear due to the variation with depth of tangential velocity. Baglin (1972) studied the onset of turbulence due to this shear. For incompressible fluids the condition for a shear flow to become turbulent is that the Reynolds number Re becomes larger than a critical value of the order several thousands

$$Re = \frac{V\theta l}{\nu} > R_c . \tag{10}$$

In stars, ν , the viscosity is dominated by the radiative viscosity in the physical conditions we are interested in.

$$\nu = \nu_o \ T^{7.5} \ \rho^{-3} \qquad \text{(with a Kramer opacity law).}$$

It is generally very small so that Reynolds numbers are high and shear instabilities are favoured.

Using for the approximation of the horizontal velocity field, the same as in eq 8, one defines the characteristic length of the shear

$$l = \left(\frac{\partial}{\partial r} \ln V_\theta\right)^{-1}.$$

And the condition for the onset of turbulence is then

$$T^{-6.5} \rho^2 V^2 > \text{Const.} \qquad (11)$$

which defines a critical velocity above which the flow becomes turbulent.

Using classical parameters for A stars Baglin (1972) obtained $V_{crit} \simeq 50$ km s^{-1} for turbulence below the second helium convective zone. However this result suffers quite important criticism.

This instability will determine the critical velocity of mixing only if it is the only one working at low rotational velocities.

3 Differential rotation

Zahn (1974) has shown that differential rotation in itself should be a very powerful agent of destabilization.

At least in the slow rotation regime the meridional circulation velocity field induces a differential rotation as a consequence of the conservation of angular momentum. In this process a gradient of angular velocity is built, for which the Reynold number becomes very rapidly larger than the Reynolds number associated to the shear.

$$Re_{diff\ rot} = Re_{shear} \times \Omega \Delta t \qquad (12)$$

over a time Δt.

This shows that the destabilization due to the building of differential rotation appears rapidly. However a detailed treatment of the diffusion of angular momentum is badly needed to settle the point and make quantitative estimates.

In gravitating media horizontal and vertical motions have different behaviours. Vertical motions are helped by the density stratification and they are submitted to the Richardson criterion of instability. However, taking into account the thermal diffusion, this criterion should be modified (as shown by Zahn 1974) strengthening the stability in the vertical direction

$$\left(\omega \frac{\partial \Omega}{\partial \omega}\right)^2 \gtrsim \sigma \ Rc \ \frac{g}{Hp} \left(\nabla_{ad} - \nabla_{rad}\right) \qquad (13)$$

ω is the distance to the axis of rotation
σ the Prandtl number ratio viscosity over thermal conductivity.
Then, the following picture emerges.

Let us define two critical "equatorial velocities".

V_1 corresponding to the onset of instability of the meridional circulation or differential rotation induced by it through the Reynolds criterion.

V_2 corresponding to the instability in the vertical direction.

In the very low rotation regime $V < V_1$ the outerlayers are stable. Diffusion proceeds freely with the time scale computed precedingly.

In the high velocity regime $V > V_2$ a strong 3 dimensional turbulence sets in which mixes very rapidly the medium. This will be the regime of the variables.

In the intermediate regime $V_1 < V < V_2$ turbulence in the horizontal direction settles. It uniformizes the angular momentum over spheres. As motions are mostly horizontal they should not help diffusion.

3.4 Turbulent diffusion

Schatzman (1969) has shown that even in slowly rotating stars stellar spin-down or meridional circulation can produce a small turbulence sufficient to minimize the sorting. However in his model radiation pressure was not included and all estimates were made at the bottom of the second convective zone, which makes the numerical results difficult to use.

The "horizontal turbulence" which probably exists in the domain $V_1 < V < V_2$ might also act as a "diffusion barrier". However no estimate exists up to now of the corresponding structure of the motions.

III Discussion.

Let us list now the main difficulties encountered in this problem.

1- From the theoretical point of view, though we can give a rough sketch of the probable processes acting an exact treatment is far from our knowledge. The correct description of the hydrodynamical behaviour of the gas of a rotating star cannot be reached without a complete treatment of the three dimensional problem. Though several instability criteria can be derived the state of motion is unknown.

2- Fast rotators with Am abundances. Except if there is a strong agent which forces solid body rotation it is difficult to understand how instabilities could not develop. As usual a strong magnetic field can be invoked.

3- The slowly rotating variables. To interpret them one needs to find a process of mixing. Until we have an idea of the physical difference between these slow rotating variables and the corresponding slow rotating Am, it is difficult to attribute the mixing to a particular process, though we can imagine that other instabilities than those which have been studied here can exist.

These two last points probably mean that the situation is more than one parameter dependent and that other processes perturb the interplay between sorting and mixing, though up to now they are not known.

4- The situation in the giant domain.

The abundances anomalies found here do not seem to be too difficult to understand, as the disappearance of the underabundance of Sc (or Ca) can be attributed to a slight variation of the physical parameters of the region of sorting.

However the fact that "constant" as well as "pulsators" seem to have the same kind of abundances-anomalies is difficult to understand.

The deepening of the ionization zones during the evolution from the main sequence to the giant branch should lengthen the diffusion times scales. If the diffusion time scale becomes larger than the evolutionary one (which is becoming shorter and shorter) the abundances could be only the fossil remnants of the abundances on the main sequence.

The interpretation of all these facts are to my point of view a very interesting field of investigation which will give us new clues to the hydrodynamical behaviour of the outerlayers of the stars.

References

Abt, H.A. 1965, Astrophys. J. Suppl. 11, 429

Abt, H.A., Moyd, K.I. 1973, Astrophys. J. 182, 809

Aller, L.H., Chapman, S. 1960, Astrophys. J. 132, 461

Babcock, H.W. 1958, Astrophys. J. Suppl. 3, 141

Baglin, A. 1972, Astron. and Astrophys. 19, 45

Baglin, A. , Breger, M., Chevalier, C., Hauck, B., Le Contel, J.M., Sareyan, J.P., Valtier, J.C. 1972, Astron. and Astrophys. 23, 221

Bartolini, C., Gilli, F., Parmeggiani, G. 1972, I.A.U. Inf. Bull. Com. 27 n°704

Bessel, M.S., Eggen, O.J. 1972, Publ. Astr. Soc. Pac. 84, 72

Breger, M. 1970, Astrophys. J. 162, 597

Breger, M. 1974a, Symp. IAU n° 59 Stellar Instabilities and Evolution Ledoux et al (ed), D. Reidel Publishing Company.

Breger, M. 1974b, Symp. IAU n° 67 Moscou

Breger, M., Maitzen, H.M., Cowley, A.P. 1972, Publ. Astron. Soc. Pac. 84, 443

Cameron, R.C. 1967 in Magnetic and Related Stars, p. 567, Monobook Corporation, Baltimore

Chapman, S., Cowling, T.G. 1960, Mathematical Theory of non uniform Gases, Cambridge University Press.

Conti, P. 1970, Publ. Astron. Soc. Pac. 82, 781

Cox, A.N., King, D.S., Tabor, J.E. 1973, Astrophys. J. 184, 201

Deutsch, A.J. 1967, in Magnetic and Related Stars, p. 181, Monobook Corporation, Baltimore.

Ishikawa, M. 1973, Publ. Astron. Soc. Japan 25, 111

Kobayashi, M., Osaki, Y. 1973, Publ. Astron. Soc. Japan 25, 495

Kohl, K. 1964, Zeit. für Astrofys. 60, 115

Mestel, L. 1965, in Stars and Stellar Systems, vol. 8, Mc Laughlin and Aller ad. (Univ. Chicago Press).

Mendoza, E.E.V., Gonzales, S.F.P. 1974, Revisto Mexicana de A. y A. 1, 67

Michaud, G. 1970, Astrophys. J. 160, 641

Schatzman, E. 1969, Astron. and Astrophys. 3, 331

Slettebak, A. 1955, Astrophys. J. 121, 653

Smith, M.A. 1971, Astron. and Astrophys. 11, 325

Smith, M.A. 1972a, Astron. and Astrophys. 19, 312

Smith, M.A. 1972b, Astrophys. J. 175, 765

Smith, M.A. 1973, Astrophys. J. Suppl. 25, 277

Smith, M.A. 1974, Astrophys. J. 189, 101

Toomre, Y., Zahn, J.P., Latour, J., Spiegel, E.A. 1975, preprint.

Vauclair, G., Vauclair, S., Pamjatnikh, A. 1974, Astron. and Astrophys. 31, 63

Zahn, J.P. 1972, C.R.A.S. 274, 1443

Zahn, J.P. 1974, in IAU Symp. n° 59 Stellar Instabilities and Evolution p. 185, Ledoux et al ed. Reidel Publishing Company.

Discussion to the paper of BAGLIN

FITCH: If you need a mixing mechanism to explain the observed pulsation of some slow rotators, would tidally driven mixing currents help?

BAGLIN: Even in the case of 14 Aur, probably the perturbation is too small to be significant.

MOLNAR: You mentioned a problem with the diffusion mechanism is that it is too efficient. That is, overabundances should be larger in your stars than they are observed to be.

One way to quench the diffusion mechanism (in middle A and late A stars) is the effect of increasing ultraviolet opacities. For these stars, a small metal overabundance quickly cuts off the ultraviolet radiation, halting the upward diffusion of the heavy elements. Because the ultraviolet radiation is the main driving force in a stellar atmosphere, future diffusion models should include the effects of increasing metal opacities.

BARTOLINI: I would like to show the results of my observations of 32 Vir, taken at the Bologna Observatory during the three consecutive nights of 20, 21, 22 March 1973. At the beginning of the first night the star was practically constant, but after $0^d.3$ it was clearly pulsating. During the second night the star was pulsating with increasing amplitude. Pulsations were present at the beginning of the third night but stopped at the end. From these observations it is possible to deduce two times of minimum amplitude; their difference of $1^d.8$ is a multiple of the beat period. I think the beat period should be $P_b \simeq 0^d.6$, while the fundamental period is $P_o = 0^d.076$.

FROLOV: I would like to point out that there is a δ Scuti star which is also an Am star, namely BS 3588 = FZ Vel.

BAGLIN: Before making a definite statement on such a star, it has to be studied in detail. Among A stars, binaries are frequent, and often difficult to distinguish spectroscopically because they are generally associated with another A star, as for example 32 Vir. On the other hand, as I have shown, the luminosity class is an important parameter.

Multiple Periodic Variable Stars
IAU Colloquium No.29, Budapest, 1975

MULTIPLE PERIODICITIES IN CATACLYSMIC VARIABLE AND WHITE DWARF STARS

Brian Warner

Department of Astronomy, University of Cape Town

1. Dwarf Novae

The normal dwarf nova (DN) outburst is characterized by a rapid rise, a brief stay near maximum and a slower decline to minimum light; the whole sequence lasts in general for 4 to 6 days, but can take as long as 10 or 12 days in some stars. While not strictly periodic, the DN outbursts are quasi-periodic in the sense that the mean periods derived from intervals sufficient to contain at least 50 outbursts (i.e. a few years for the shortest period systems, up to 30 years for longer period systems) remain sensibly constant. It should be added, however, that in the case of U Gem there is evidence for an increase in its recurrence period which may indicate some evolutionary change: for the first 50 years of observation of U Gem the mean period was 96.5 days and for the second 50 years it was 107.6 days (Mayall, 1957); during the past four years the mean interval has been 291 days (with only one outburst in 1974).

The shortest period systems - those with normal recurrence periods $P_n \lesssim 30$ days - can almost all be divided into two subgroups. These are the well-known Z Cam subgroup in which standstills occur at apparently random intervals, and a subgroup which we propose to call the SU UMa stars which show supermaxima. In supermaxima, the star rises initially to a peak brightness about 0.5 magnitude brighter than the normal maximum, and then (often with large fluctuations) hovers around normal maximum brightness for between 10 and 20 days (corresponding to factors ~ 4 times longer than the duration of normal maxima) before declining rapidly to minimum brightness.

We suspect that those stars with $P_n < 30$ days that are not yet known to fall into either the Z Cam or SU UMa subgroups are poorly observed and that further studies may well result in all such objects being classifiable into the two subgroups. The Z Cam or SU UMa characteristic occasionally occurs in stars with $P_n > 30$ days; for example BS Cep is a Z Cam star with $P_n \sim 40$ days and Z Cha is a SU UMa star with $P_n \sim 96$ days.

Bateson (1975a) has pointed out that the shorter the mean interval between supermaxima, the more regular and predictable they become. For these regular supermaxima we can therefore define a "super-period" P_s, and we can classify such stars as multiple quasi-periodic systems. Information on SU UMa stars (mostly drawn from Circulars of the Variable Star Section of the Royal Astronomical Society of New Zealand and from the Variable Star Section reports appearing in the Journal of the British Astronomical Association, plus private correspondence with the present Directors: F. M. Bateson and J. E. Isles) is given in Table 1.

Table 1

The SU UMa Stars

Star	P_n	Supermaxima
SU UMa	13	$P_s = 180$ days
AY Lyr	31	$P_s = 200$ days
VW Hyi	28	$P_s = 180$ days
WX Hyi	29	$P_s = 165$ days
Z Cha	96	$P_s = 313$ days
V436 Cen	25:	Random, mean interval ~ 630 days
AT Ara	70	Random, mean interval ~ 500 days
CU Vel	139	Scattered
OY Car	188	$P_s \sim 300$ days

There are four stars with $P_s = 180 \pm 20$ days and these have the shortest periods and most regular behaviour. Bateson (1975b) finds that $P_s = 179.6 \pm 12.2$ days for VW Hyi, which means that supermaxima are predictable to at least the same accuracy as the most regular Mira variables. In fact, the scatter may be even less than that quoted: there is some ambiguity as to when maximum brightness occurs in the supermaxima - they stay bright for many days and can have large fluctuations. Timings made on the rising branch (e.g. passing through a particular magnitude) should be investigated to see if they improve the predictability. However, some supermaxima (in VW Hyi) are known to occur soon after a normal outburst, or even while one is in progress. This confusion must be eliminated before timings of the rising branch are analysed.

A statistical analysis of possible relationships between normal and super-maxima needs to be made - preferably in several objects. The occasional occurrence of supermaxima that begin while a normal maximum is underway suggests that the supermaxima may be triggered by normal outbursts. At the moment we have no idea of the cause of the multiple periodicity in the SU UMa stars. It has been shown (Warner 1975) that the time-averaged energy radiated in the visible during the normal outburst cycle and the supermaximum cycle is roughly the same implying that the same energy source is being tapped with the same efficiency in the two cases. The light curves should be analysed to see if a supermaximum has any significant effect on subsequent normal maxima; if this is the case then this would afford more evidence that the two periodicities are modulations of the same energy-generating process.

At maxima of VW Hyi, the colours are B-V = -0.02, U-B = - 0.75 (Marino and Walker 1973, Eggen 1968) and during supermaximum they are B-V = -0.08, U-B = -0.75 (Vogt 1975a). Spectra taken during a supermaximum of VW Hyi (Vogt 1975b) are similar to those obtained during normal maxima of other systems (Joy 1960). There do not, therefore, seem to be any large differences in the physical properties of dwarf novae at normal and supermaxima.

Of the stars listed in Table 1, only three have orbital periods determined; these are VW Hyi (P = 107 mins), Z Cha (P = 107 mins) and V436 Cen (P = 92 mins). It may be significant that these three stars are ultra-short period binaries: this may be necessary for the SU UMa phenomenon. None of the Z Cam stars is known to have such short orbital periods; those known are EM Cyg ($P = 7^h 00^m$), Z Cam ($P = 6^h 56^m$), RX And ($P = 5^h 05^m$), WW Cet ($P = 3^h 50^m$) and CN Ori ($P \sim 6^h$). Furthermore, the other ultra-short period systems all have peculiarities: WZ Sge (P = 82 mins) is a dwarf nova with a recurrence period of 33 years, EX Hya (P = 99 mins) has outbursts every $\sim$465 days but many more frequent low amplitude outbursts (the larger outbursts associated with the 465 day timescale do not have the characteristics of supermaxima but nevertheless may be loosely related) and VV Pup (P = 100 mins) which has never been seen to outburst but may be like WZ Sge with a much longer recurrence timescale.

2. Rapid Oscillations in Cataclysmic Variable Stars

The rapid oscillations of brightness discovered in cataclysmic variable stars, especially in dwarf novae during outbursts (Warner and Robinson 1972a), have been seen in a total of fourteen stars (Warner 1975). Among these, some evidence for multiple periodicities exists.

Firstly, however, it should be stated that the original claim for the existence of multiple discreet periodicities in dwarf novae and for transitions between these, has largely been discounted by the more detailed analysis of Warner and Brickhill (1975). Apparent multiple periodicities can arise in periodogram analyses if there is amplitude modulation of the periodic signal, and the characteristic timescale of the modulation is comparable with the data length used for computation of the periodograms. Under these circumstances, large side lobes can appear to the principal peak of a periodogram and these give the impression of multiple periodicities. As a result, the various apparent multiplicity of periodicities that occur in some of the periodograms of the cataclysmic variables must be treated with reservation.

Several types of periodic modulation of the rapid oscillations in dwarf novae are, however, present. The orbital period is seen to modulate the rapid oscillations in VW Hyi and V436 Cen (Warner and Brickhill 1975) with a total period range ~5 percent. A smaller modulation, displayed as a phase shift around the orbit (excluding the region near eclipses) is seen in DQ Her (Warner et al 1972) and UX UMa (Nather and Robinson 1974). Still unexplained, these modulations place some restrictions on what mechanisms may be responsible for the rapid oscillations.

Longer term modulations of the rapid oscillations may exist. The oscillations seen in UX UMa (Warner and Nather 1972a, Nather and Robinson 1974) have shown a range of 28.55 to 30.03 secs, which may be cyclical with a period of weeks. Observations of CD -42°14462 (Warner 1973, Hesser et al 1974, Warner unpublished) range from 29.1 to 32.0 secs and could be showing the same phenomenon as in UX UMa. Whether these longer term variations will be explicable in terms of orbital motion due to the presence of a third body remains to be seen.

A 413 sec modulation of the 30 sec oscillations present during a normal maximum of VW Hyi has been found (Warner and Brickhill 1975). This period is of the right order of magnitude to be attributed to rotation of an obscuring cloud near the outer edge of the accretion disc that surrounds the white dwarf primary in this system. It could also be identified with oscillatory motion of the disc as a whole, periodically obscuring the oscillating source near the centre of the disc.

As can be seen, the cataclysmic variables possess a multitude of periodic phenomena. To summarise the properties of just two: UX UMa shows a 29 year variation of its 4^h43^m orbital period (Nather and Robinson 1974) and has an orbital modulation and a possible modulation with a period of weeks of its $\sim$30 sec oscillations. VW Hyi shows outburst quasi periodicities on timescales of 180 days and 28 days, has a 1^h47^m orbital period which can modulate its $\sim$30 sec oscillations and has also on one occasion shown a 413 sec periodicity.

3. Variable White Dwarfs

Generalities

The first of the variable white dwarfs, HL Tau-76 (V411 Tau) was discovered in 1964 by Landolt (1968) while conducting a general photometric program. At about the same time, a special search was started by the Princeton group using power spectrum techniques to look for coherent periodicities in white dwarfs in the period range 1 - 1000 secs. This initial survey was not fruitful which served to show that variability is a rare occurrence among white dwarfs picked at random (Lawrence et al 1967, Hesser et al 1969). Extension of this survey by observations made at Cerro Tololo resulted in the discovery of two low-amplitude white dwarfs out of a total of 23 observed; a summary of this program has been given by Hesser and Lasker (1972). The two new variables, G44-32 (CY Leo; Lasker and Hesser 1969) and R548 (ZZ Cet; Lasker and Hesser 1971) have amplitudes of only $\sim$0.01 - 0.02 mag, whereas HL Tau-76 has $\sim$0.3 mag variations. Smak (1967) discovered variability in HZ 29 (AM CVn), a star with a spectrum resembling a DB white dwarf, but this object has subsequently been considered to be an ultra-short period binary (Warner and Robinson 1972b).

During the past two years, discoveries of variable white dwarfs have increased dramatically. Partly this is due to the larger number of groups searching for variability, but it is more a consequence of noting (Lasker and Hesser 1971) that HL Tau-76, G44-32 and R548 lie close together in the two-colour plane near B-V = 0.25, U-B = -0.55, with the resulting concentration on stars in that region. Richer and Ulrych (1974) claimed variability in G117-B15A and G169-34, confirmation of variability in the former has been given but G169-34 appears to have constant brightness (McGraw and Robinson 1975a). Schulov and Kopatskaya (1973) discovered that G29-38 is variable, which has been confirmed and a further star, G38-29, added by McGraw and Robinson (1975b). Three further objects, GD99, R808 and G207-9 have recently been found (McGraw and Robinson 1975 a, c). We therefore have a total of nine known variable white dwarfs; these are listed in Table 2. It is a significant fact that all of these are multi-periodic.

The first four white dwarf variables (V411 Tau, ZZ Cet, CY Leo and AM CVn) have been given the new classification of ZZ Cet stars in the Second Supplement to the General Catalogue of Variable Stars. To judge from the amplitudes of variability listed in Table 2, there may be two distinct types, with ZZ Cet unfortunately representing the minority group. We propose a more general classification, the DV stars, which (d.v.) will suffice to cover all possible classes of variable white dwarfs and allow of subgroups such as ZZ Cet or V411 Tau stars.

Table 2

DV Stars

EG	Name	Spectrum	B-V	U-B	Amplitude (mags)	Periods (secs)
10	R548	DA	0.20	-0.54	0.01	212.86, 273.0
34	G38-29	DAs	0.16	-0.53	0.20	925, 1022
65	G117-B15A	DA	0.20	-0.56	0.08	216, 272, 308
72	G44-32	DC	0.29	-0.58	0.02	1638, 822, 600
115	R808	DA	0.17	-0.56	0.15	833 + others
127	G207-9	DAn	0.17	-0.60		
159	G29-38	DA	0.20	-0.65	0.23	613, 671, 816, 930 1000
219	GD99	DA	0.19	-0.59	0.10	229 + others
265	HL Tau-76	DA	0.20	-0.50	0.30	746, 494

Individual Stars

R548 (ZZ Cet) Lasker and Hesser (1971) found two nearly sinusoidal variations with periods of 213 and 273 secs, whose relative amplitudes vary on a time scale of hours, both being up to $0^{m}.01$. Hesser and Lasker (1972) have additional photometry suggesting that at least the 213 sec oscillation is probably coherent on a long timescale. More extensive observations are required on this star.

G44-32 (CY Leo) This star also has low and variable amplitude oscillations (Lasker and Hesser 1969). Two periods and a probable harmonic are present.

HL Tau-76 (V411 Tau) Landolt's (1968) discovery observations were made in poor photometric conditions but a periodogram analysis showed the presence of a periodicity near 12.5 mins. The photometry by Warner and Nather (1972b) clarified the nature of the light curve. The brightness variations have a range up to $0^{m}.3$, are highly non-sinusoidal and exhibit a large range of pulse shapes. Further observations were made by Fitch (1973). Power spectrum analyses of these observations (Warner and Robinson 1972a; Page 1972; Fitch 1973) show a fairly stable spectrum of oscillation components including a rich series of harmonics and cross components of the principal frequencies. We omit further discussion as further details may be found in the contribution by Desikachavy and Tomazewski later in this Colloquium.

G29-38 Observations by McGraw and Robinson (1975b) have extended those of the discoverers (Shulov and Kopatskaya 1973). The light curve is strikingly similar to that of HL Tau-76. The power spectrum shows a qualitatively similar pattern from night to night, but there are small variations in the peaks and their relative amplitudes which can occur on a timescale of hours. There is a nearly equal spacing of the major spectral peaks in the frequency plane.

G38-29 This star also has a light curve like that of HL Tau-76 (McGraw and Robinson 1975b). Power spectra of different runs again show some instability in the periods and relative amplitudes.

G117-B15A Richer and Ulrych (1974) obtained rather poor-quality observational data on this star and an analysis by the Maximum Entropy Method indicated a period of 1311 secs with an amplitude of 0.01 mag. However, observations by McGraw and Robinson (1975a) show G117-B15A to have an almost sinusoidal light curve with period 216 secs and amplitude 0.08 mag. Two smaller peaks in the power spectra are found at 272 and 308 secs with amplitudes ~ 0.001 mag. Power spectra of different runs are almost identical.

R808 With a light curve similar to that of HL Tau-76, R808 shows complex structure in its power spectrum, with many peaks in the range 770-1250 secs (McGraw and Robinson 1975a). The power spectra are unstable.

GD99 This star has a very irregular light curve and at times may be quiescent for intervals of up to an hour (McGraw and Robinson 1975a). Only one peak, at 229 secs, in the power spectrum remains constant in frequency; other peaks vary in frequency and amplitude.

G207-9 Recently discovered by Robinson (McGraw and Robinson 1975c) this object is a low-amplitude variable.

Discussion

Several features of significance to the interpretation of the variability in white dwarf stars emerge from these studies. We note that, with the exception of G44-32 (which is faint and for which further spectra should be obtained) all DV stars are of spectral type DA. From their colours, they lie near the cool end of the observed DA sequence. The range of colours is small; although this may be affected by observational selection, a survey of white dwarfs lying outside the two-colour limits of the stars listed in Table 2 has failed to find any variables (Hesser and Lasker 1972, McGraw and Robinson 1975c). Not all of the stars lying within the range of colours of the DV stars are variable. This may be partly due to photometric errors scattering some non variables into the area of the variables, but it may also imply differences in internal structure of the DV and non-variable stars. The region of the two-colour diagram that the DV stars occupy, when transformed to the M_{bol} - log T_e diagram with the aid of model atmosphere calibrations, lies in the same region as that where calculated white dwarf evolutionary sequences (and DA stars with known distances) pass through an extension of the Cepheid instability strip (Vauclair 1971).

The low amplitude ($\Delta m < 0.10$ mag) DV stars possess sinusoidal light variations with constant periods. The larger amplitude systems have highly non-sinusoidal variations, with cross-terms in their frequency spectra, and have unstable periods. The long periods of the DV stars excludes radial oscillations and suggests non-radial modes of oscillation (Brickhill 1975). The presence of many closely-spaced eigenvalues supports this contention. Instability of the eigenfrequencies may be a result of interaction between the g-mode oscillations and the convective layer: the g-mode eigenfunctions corresponding to pulsations having periods of hundreds of seconds will have large amplitude in the convective region of the envelope.

Further theoretical work is clearly required. Some simplification may be offered from the fact that the DV stars have DA atmospheres which are probably the best understood. However, for a full understanding of the light curves it will probably be necessary to develop the non-linear, non-adiabatic, non-radial theory, with additional complications such as interaction with convection. In the meantime, extensions of the survey for DV stars, especially to the southern hemisphere, should result in many further discoveries, the systematics of which may provide some further clues to the nature of the variability.

References

Bateson, F.M., 1975a. Private communication.

Bateson, F.M., 1975b. Roy. Astron. Soc. New Zealand, Var. Star Sect. Circ. M75/5.

Brickhill, A.J., 1975. Monthly Notices Roy. Astron. Soc. 170, 405.

Eggen, O.J., 1968. Quart. J. Roy. Astron. Soc. 9, 332.

Fitch, W.S., 1973. Astrophys. J. Letters 181, L95.

Hesser, J.E. and Lasker, B.M., 1972. IAU Colloq. No. 15, p. 160.

Hesser, J.E., Ostriker, J.P. and Lawrence, G.M., 1969. Astrophys. J. 155, 919.

Hesser, J.E., Lasker, B.M. and Osmer, P.S., 1974. Astrophys. J. 189, 315.

Joy, A.H., 1960. in Stellar Atmospheres, ed. Greenstein, Univ. Chicago Press, p. 653.

Landolt, A.U., 1968. Astrophys. J. 153, 151.

Lasker, B.M. and Hesser, J.E., 1969. Astrophys. J. Letters 158, L171.

Lasker, B.M. and Hesser, J.E., 1971. Astrophys. J. Letters 163, L89.

Lawrence, G.M., Ostriker, J.P. and Hesser, J.E., 1967. Astrophys. J. Letters 148, 161.

Marino, B.F. and Walker, W.S.G., 1974. Inform. Bull. Var. Stars No. 864.

Mayall, M.W., 1957. J. Roy. Astron. Soc. Can. 51, 165.

McGraw, J.T. and Robinson, E.L., 1975a. Preprint.

McGraw, J.T. and Robinson, E.L., 1975b. Preprint.

McGraw, J.T. and Robinson, E.L., 1975c. Private communication.

Nather, R.E. and Robinson, E.L., 1974. Astrophys. J. 190, 637.

Page, C.G., 1972. Monthly Notices Roy. Astron. Soc. 159, 25P.

Richer, H.B. and Ulrych, T.J., 1974. Astrophys. J. 192, 719.

Schulov, O.S. and Kopatskaya, E.N., 1973. Astrophysica 10, 117.

Smak, J., 1967. Acta Astron. 17, 255.

Vauclair, G., 1971. Astrophys. Letters 9, 161.

Vogt, N., 1975a. Astron. Astrophys. 36, 369.

Vogt, N., 1975b. IAU Symp. No. 73. In press.

Warner, B., 1973. Monthly Notices Roy. Astron. Soc. 163, 25P.

Warner, B., 1975. IAU Symp. No. 73. In press.

Warner, B. and Brickhill, A.J., 1975. Monthly Notices Roy. Astron. Soc. In press.

Warner, B. and Nather, R.E., 1972a. Monthly Notices Roy. Astron. Soc. 159, 429.

Warner, B. and Nather, R.E., 1972b. Monthly Notices Roy. Astron. Soc. 156, 1.

Warner, B. and Robinson, E.L., 1972a. Nature Phys. Sci. 239, 2.

Warner, B. and Robinson, E.L., 1972b. Monthly Notices Roy. Astron. Soc. 159, 101.

Warner, B., Peters, W.L., Hubbard, W.B. and Nather, R.E., 1972. Monthly Notices Roy. Astron. Soc. 159, 321.

Discussion to the paper of WARNER

COX: I had thought that the case of DQ Her was settled in favor of rotation by Kemp and coworkers. Yet you still say the 71 second variation is pulsation rather than a 142 second rotation. Could you comment?

WARNER: Power spectra analysis of DQ Her by Nather and Kiplinger show no power at 142 second period. This throws some doubt on the various claims for a 142 second variation in circular polarization of DQ Her, on which the conclusion that rotation is favored was based. This does not, of course, exclude rotation as a still possible mechanism.

BATESON: Regarding Warner's reference to my work, I would like to add some comments on observations in New Zealand. In 1954, we commenced to observe U Gem stars in the southern sky. With a wide spread of observers in latitude (S5° to S46°) and a very wide spread in longitude in southern countries, the best observed have resulted in such stars being covered on most nights each year. Two decades of observations, much already published, now enable a discussion of the gross behavior of such stars to be published, and these will be in a series of memoirs, the first due late in 1975. In 1970, photoelectric observations were commenced at the Auckland Observatory both during outbursts of U Gem stars and during minima. Now three groups have photoelectric equipment. In 1974 at Carter Observatory a program commenced of observation of U Gem stars fainter than magnitude 14 at maximum, using photographic photometry. This will result in finding and detailed charts for these faint members as part of the Chart Series of the V.S.S., R.A.S.N.Z. Charts for most of the brighter stars have already been published. Observational results appear in the Publications, V.S.S., R.A.S.N.Z. During the course of the present meeting, Janet Mattei and myself have agreed on a cooperation program to avoid duplication of observations between the A.A.V.S.O. and V.S.S., R.A.S.N.Z.

BATESON: Were your observations of RR Pic of the shell or of the remnant?

WARNER: The remnant.

MATTEI: Gordon in 1950, Walker and others later have reported irregular variations within seconds, minutes, and hours, during minima of SS Cyg. Would you comment if these irregular variations are of the same nature as those variations you have observed?

WARNER: The photometric observations of those workers, and unpublished observations that I have made, show that in SS Cyg there is rapid flickering activity. Our analyses show that this flickering is nonperiodic, i.e., it is random or stochastic. We have not, however, observed SS Cyg during an outburst.

WOLFF: You indicated that the variable white dwarfs occupy a restricted range in colors. If you observe a sample of DA white dwarfs with these colors, what fraction have detactable variations?

WARNER: One third.

BATH: Regarding Cox's question, a rotation model of DQ Her 71 second oscillation does not necessarily lead to significant power in a periodogram analysis at 142 seconds. Even a non-sinusoidal oscillation could give negligible power at the 142 second period, but spread the power over _many_ higher harmonics. If it is indeed a pure sine wave, then that is no argument at all in favor of pulsation as opposed to rotation as an explanation.

With reference to Dr. Cox's point, there has been frequent use of the word pulsation in the discussions, to describe periodicities observed in many systems. To a theoretician, pulsation is a heavily loaded word. Could I plead that observers do not use the word pulsation to describe observed phenomena that may have nothing to do with stellar pulsation.

Multiple Periodic Variable Stars
IAU Colloquium No.29, Budapest, 1975

NON-RADIAL OSCILLATIONS OF DEGENERATE DWARFS: A THEORETICAL REVIEW

H.M. Van Horn

Department of Physics and Astronomy
and
C.E. Kenneth Mees Observatory
University of Rochester
Rochester, N.Y. 14627

Abstract. The current status of the theory of non-radial oscillations of white dwarfs and hydrogen shell-burning degenerate stars, in the linear, quasi-adiabatic approximation, is reviewed. Relevant aspects of the thermal structure of such stars are summarized, and several problem areas requiring further investigation are identified. Among the most interesting and important of these are the question of the existence of g^- modes in white dwarfs; the nature of the excitation mechanisms for non-radial oscillations; the possibility of tidal excitation; the problem of mode-coupling between pulsations and convection; and the question of the effects of a solid, crystallizing core.

1. Introduction

Recent observations, as reviewed by Warner (1975) at this conference, have shown the existence of two classes of degenerate stars exhibiting multiperiodic variability. One class is comprised of Z Cam stars, which are a subgroup of the dwarf novae. These stars display optical periodicities in the 20-30 second range, but only near the peak of an outburst. The oscillation periods are too long for radial pulsations of stars of ~ $1.2M_{\odot}$, which is typical of the dwarf novae (Warner 1973), unless the Z Cam stars are very close indeed to the limiting mass for their internal chemical composition (cf. Warner 1974). The fundamental mode radial pulsation periods for degenerate dwarfs are Π_o ~ 2 to 5 seconds for stars in the mass range ~ 1.0 to $1.4M_{\odot}$, depending upon the composition (Wheeler, Hansen, and Cox 1968; Faulkner and Gribbin 1968). For this reason, and because of the existence of multiple periodicities with small frequency separations (uncharacteristic of radial modes, but understandable in terms of rotation-splitting of the non-radial modes); because of "mode-hopping" in these stars; and because the oscillation period is sometimes observed to change significantly during the course of an outburst, Warner and Robinson (1972) and Chanmugam (1972) have suggested that the high-frequency variability in these stars is associated with non-radial "g-mode" oscillations of the degenerate dwarf.

The second class of degenerate variable stars consists of the variable white dwarfs. This group is much less homogeneous than the Z Cam stars and includes i) suspected or known binaries, ii) a variety of spectral types, and iii) effective temperatures ranging from ~ 9000°K to more than 20,000°K. (There is, however, evidence for some concentration of variable white dwarfs with colors near B - V ≈ +0.25, U - B ≃ -0.55: Lasker and Hesser 1971, Richer and Ulrych 1974). The observed periods in this group, most of which exhibit multiple periodicities, range from 29.08 and 30.15 seconds in CD-42° 14462 (Hesser, Lasker, and Osmer 1974) to 1311 seconds in EG65 (Richer and Ulrych 1974). Since the absolute magnitude of EG65 is known, a photometric estimate of the mass is possible, giving M ~ $1.0M_{\odot}$ (Richer and Ulrych 1974). The observed periods are thus again *much* too long to be associated with radial pulsations, and it has been suggested that these are also g-mode oscillations (Chanmugam 1972; Warner and Robinson 1972; Hesser, Lasker, and Osmer 1974).

It is accordingly the purpose of this review to summarize the current state of our theoretical understanding of non-radial oscillations in degenerate dwarfs. An excellent recent discussion of this problem is contained in the review paper of Ledoux (1974), and I shall mainly attempt to supplement his summary and bring it up to date. To place the discussion in context, those aspects of white dwarf structure which are relevant to the non-radial oscillation problem are first summarized briefly in Section 2. In Section 3 the linear quasi-adiabatic theory of non-radial oscillations is briefly reviewed in order to establish notation and to emphasize the role of the non-adiabatic terms. Specific calculations are reviewed in Section 4 and we conclude in Section 5 with a summary of the current status of the theoretical situation and an identification of certain aspects of the problem that merit immediate attention.

2. The Physical Structure of Degenerate Dwarfs

White dwarfs are well-known to be stars which are supported against their self-gravitational attraction primarily by the pressure of highly degenerate electrons (cf. Chandrasekhar 1935, 1957; Hamada and Salpeter 1961). White dwarf masses are typically in the range 0.5 to $1.0M_\odot$, and radii are $R \sim 10^{-2}R_\odot$. Fundamental mode radial oscillation periods are thus expected to be of order $\Pi_o \sim (G\bar{\rho})^{-1/2} \sim 10$ seconds. Because of the high degeneracy, electron conduction is extremely efficient, and the degenerate core of a white dwarf is very nearly isothermal. The luminosity is supplied from the thermal content of the interior (Mestel 1952):

$$T\frac{ds}{dt} \approx C_v\frac{dT}{dt} \neq 0. \tag{1}$$

In such circumstances a star is said to be in "thermal imbalance" (radiative losses not balanced by nuclear energy production), and this has a bearing on the vibrational stability of the star.

The core of a white dwarf probably consists primarily of a mixture of carbon and oxygen, and at sufficiently low temperatures the ions may crystallize onto a regular lattice structure. The best current estimate for the crystallization temperature is given by:

$$\Gamma = 2.28 \frac{Z^2}{A^{1/3}} \frac{(\rho/10^6 g\ cm^{-3})^{1/3}}{(T/10^7 {}^\circ K)} \approx 160 \qquad (2)$$

where the ratio of Coulomb to thermal energies Γ is $\approx$ 160 at the point of crystallization of the plasma (Hansen 1973, Lamb 1974). Crystallization temperatures are high enough to occur at the observed luminosities of the white dwarfs ($L \sim 10^{-2}$ to 10^{-4} $L_\odot$); the cooler and more massive white dwarfs thus have a solid core within the fluid star.

Because the white dwarf core is nearly isothermal, most of the temperature drop occurs in a very thin non-degenerate surface layer. If this region is hydrogen-rich, the layer is radiative for surface temperatures $T_e \gtrsim$ 16,000°K, and the luminosity L is related to the core temperature T_c by (cf. Schwarzschild 1958, Van Horn 1971)

$$\frac{L}{L_\odot} \sim 2 \times 10^{-3} \frac{M}{M_\odot} \left(\frac{T_c}{10^7 {}^\circ K}\right)^{3.5} . \qquad (3)$$

Cooler white dwarfs and degenerate stars with He- or C-rich envelopes all have thin subsurface convection zones, however (Böhm 1968, 1969, 1970; Böhm and Cassinelli 1971a, b; Böhm and Grenfell 1972; Fontaine 1973; Baglin and Vauclair

1973; D'Antona and Mazitelli 1974, 1975; Fontaine *et al.* 1975). Since the existence of a convection zone is a necessary precondition for the existence of dynamically unstable g^- modes of non-radial oscillation, we therefore expect such modes to occur in sufficiently cool white dwarfs. Convection may also mix surface (or accreted) hydrogen deep enough for nuclear burning (which may affect the stability of white dwarf oscillations: cf. Richer and Ulrych 1974), and convective timescales are short enough so that interactions between convective motions and the non-radial oscillation modes may be important.

Some of the properties of the surface convection zones of $0.612M_{\odot}$, He-rich envelope white dwarfs, taken from the calculations of Fontaine (1973; see also Fontaine *et al.* 1974) are listed in Table 1.

Table 1

log Te	v_c (cm s^{-1})	z(cm)	$\tau=z/v_c$	logΔM_c	log T_b	$\varepsilon_b/X_H{}^2$	log T_c	Γ_c (^{12}C)
4.280	2.091 +5	2.920 +5	1.40	——	5.341	——	7.284	28.6
4.217	1.146 +5	1.544 +6	13.5	-8.037	6.072	2.543 -8	7.188	35.7
4.155	7.006 +4	3.663 +6	52.3	-6.352	6.411	4.757 -4	7.090	44.7
4.092	5.066 +4	4.933 +6	97.4	-5.622	6.527	1.272 -2	6.995	55.6
4.030	3.647 +4	4.806 +6	131.8	-5.539	6.496	9.598 -3	6.893	70.3
3.967	2.569 +4	4.199 +6	163.4	-5.607	6.427	2.954 -3	6.788	89.6
3.905	1.738 +4	3.254 +6	187.2	-5.768	6.334	4.716 -4	6.672	117.
3.842	1.222 +4	2.571 +6	210.4	-5.976	6.276	4.535 -5	6.566	153.
3.780	8.815 +3	2.112 +6	239.6	-6.217	6.108	2.728 -6	6.427	206.
3.717	6.043 +3	1.563 +6	258.6	-6.376	5.960	——	6.274	293.

The notation $1.00 + n \equiv 1.00 \times 10^n$ is employed, and all units are c.g.s. except for the mass of the convection zone, ΔM_c, which is given in solar masses.

As the effective temperature Te decreases, the central temperature T_c drops monotonically, and the value of Γ at the center ($\propto \rho_c^{1/3}/T_c$), computed assuming a carbon core, increases. Crystallization begins at the center for log $L/L_\odot \approx -3.55$. As the star cools, the extent of the partial ionization region increases inward until, at log Te $\approx$ 4.1, the base of the convection becomes degenerate, and the increasing efficiency of electron conduction pushes the zone back out toward the surface. The depth z, convection zone mass ΔM_c and temperature T_b at the base of the zone thus first increase with decreasing L and subsequently decrease. The nuclear burning rate ε_b at the convection zone base, due to the reaction $3\ {}^1H \rightarrow {}^3He + e^+ + \nu$ (Clayton 1969, p. 376) and computed assuming $X_H << 1$, thus increases sharply but peaks at relatively high Te (log Te ~ 4.1). Since both ΔMc and the maximum T_b increase somewhat on going to lower mass stars, the conditions for the excitation of white dwarf oscillations by nuclear burning of accreted hydrogen appears more favorable for hot, low mass white dwarfs than for any others.

The characteristic timescale of convection, as measured by $\tau = z/v_c$, where v_c is the maximum convective velocity, increases monotonically with decreasing temperature in these models. For the hotter models the convective velocity is high [thus requiring a relatively large super-adiabatic gradient in the outermost layers: $(\nabla-\nabla ad)_{max} \gtrsim 0.4$], and the convective timescales are comparable to the radial pulsation periods. For models cooler than ~ 13,000°K, however, τ is comparable to the non-radial g-mode oscillation periods, as we shall see.

3. Linear, Quasi-Adiabatic Theory of Non-Radial Oscillations

The linearized, Eulerian equations of fluid dynamics are (cf. Ledoux and Walraven 1958, Ledoux 1974, Cox 1974)

$$\frac{\partial \rho'}{\partial t} + \text{div}\,(\rho_o \underset{\sim}{v}') + \text{div}\,(\rho' \underset{\sim}{v}_o) = 0 \tag{4a}$$

$$\frac{\partial \underset{\sim}{v}'}{\partial t} + \frac{1}{\rho_o} \nabla p' - \frac{\rho'}{\rho_o^2} \nabla p_o + \nabla \Phi'$$
$$+\left[\frac{\partial \underset{\sim}{v}_o}{\partial t} + (\underset{\sim}{v}_o \cdot \nabla)\underset{\sim}{v}_o + (\underset{\sim}{v}_o \cdot \nabla)\underset{\sim}{v}' + (\underset{\sim}{v}' \cdot \nabla)\underset{\sim}{v}_o - \underset{\sim}{f}\right] = 0 \tag{4b}$$

$$\nabla^2 \Phi' = 4\pi G \rho' \tag{4c}$$

$$\frac{\partial p'}{\partial t} + \underset{\sim}{v}' \cdot \nabla p_o = \frac{\Gamma_1 p_o}{\rho_o}\left(\frac{\partial \rho'}{\partial t} + \underset{\sim}{v}' \cdot \nabla \rho_o\right)$$
$$+ \left[\left(-\underset{\sim}{v}_o \cdot \nabla p' + \frac{\Gamma_1 p_o}{\rho_o} \underset{\sim}{v}_o \cdot \nabla \rho'\right) + \rho_o(\Gamma_3 - 1) T_o \left(\frac{\partial s'}{\partial t} + \underset{\sim}{v}' \cdot \nabla s_o + \underset{\sim}{v}_o \cdot \nabla s'\right)\right.$$
$$\left. + \left(\frac{\rho'}{\rho_o} + \frac{\Gamma_3'}{\Gamma_3 - 1} + \frac{T'}{T_o}\right)\rho_o(\Gamma_3 - 1) T_o \frac{ds_o}{dt} + \left(\frac{\Gamma_1'}{\Gamma_1} + \frac{p'}{p_o} - \frac{\rho'}{\rho_o}\right)\frac{\Gamma_1 p_o}{\rho_o}\frac{d\rho_o}{dt}\right], \tag{4d}$$

$$\frac{\partial s'}{\partial t} + \underset{\sim}{v}_o \cdot \nabla s' + \underset{\sim}{v}' \cdot \nabla s_o = -\frac{T'}{T_o}\frac{ds_o}{dt} + \frac{1}{T_o}\left(\varepsilon' - \frac{1}{\rho_o}\,\mathrm{div}\underset{\sim}{F}' + \frac{\rho'}{\rho_o^2}\mathrm{div}\underset{\sim}{F}_o\right), \quad (4e)$$

$$\underset{\sim}{F}' = -\frac{4ac}{3}\frac{T_o^3}{\kappa_o\rho_o}\left[\left(3\frac{T'}{T_o} - \frac{\kappa'}{\kappa_o} - \frac{\rho'}{\rho_o}\right)\nabla T_o + \nabla T'\right]. \quad (4f)$$

All symbols have their usual meanings, primes denote Eulerian perturbations, $d/dt = \partial/\partial t + \underset{\sim}{v}'\cdot\nabla$, and we have assumed the flux $\underset{\sim}{F}$ to be non-convective in writing equation (4f). In addition, the body force per unit mass $\underset{\sim}{f}$ in equation (4b) is intended as the sum of all other body forces per unit mass (due, e.g., to viscosity, magnetic forces, tidal interactions, etc.). Because we wish to consider perturbations about a spherical state, we have also included the terms $\partial\underset{\sim}{v}_o/\partial t$ and $(\underset{\sim}{v}_o\cdot\nabla)\underset{\sim}{v}_o$ in this equation, even though they do not involve perturbed quantities, since they are neglected in the construction of spherical models.

Note that we have also retained all perturbation terms involving $\underset{\sim}{v}_o$ in equations (4). The order of magnitude of these terms can be seen from equation (4a). Here the first two terms are $\propto \Pi^{-1}$, where Π is the characteristic oscillation period, while the third term involving $\underset{\sim}{v}_o$ is clearly $\propto \tau^{-1}$, where τ is the timescale associated with the unperturbed velocity field. For pure gravitational contraction, $\tau \sim \tau_{KH}$, the Kelvin-Helmholtz timescale, while for rotation $\tau \sim \Omega^{-1}$(the rotation period) and τ is the convective timescale if $\underset{\sim}{v}_o$ represents convective motions.

Neglect of the $\underset{\sim}{v}_o$-terms is clearly satisfactory in the first case, but is certainly not in the others.

In equation (4d), in addition to the obvious $\underset{\sim}{v}_o$-terms, there also occur terms involving the time derivatives of the unperturbed entropy and density: ds_o/dt and $d\rho_o/dt$. The $d\rho_o/dt$-term is almost certainly negligible for degenerate dwarfs, but the ds_o/dt-term is not [cf. equation (1)]. These are the so-called "thermal imbalance" terms. Finally, the terms involving the entropy perturbation s' are the usual non-adiabatic terms which, in the quasi-linear approximation, determine the secular stability of the oscillations.

In the linear, adiabatic approximation, all terms involving $\underset{\sim}{v}_o$, $\underset{\sim}{f}$, the entropies, and the time derivatives of unperturbed quantities are neglected. The time-dependence of perturbed quantities is assumed to be of the form $e^{i\sigma t}$ and the angular dependences are assumed given by the spherical harmonic $Y_{\ell m}(\theta,\phi)$. Equations (4) then reduce to the fourth-order system (cf. Ledoux and Walraven 1958, Ledoux 1974, Cox 1974)

$$\sigma^2\zeta = \frac{\partial\chi}{\partial r} - A\,\frac{\Gamma_1 p_o}{\rho_o}\,\alpha \quad , \tag{5a}$$

$$-\frac{\Gamma_1 p_o}{\rho_o}\,\alpha = \chi - \Phi' + \zeta\frac{1}{\rho_o}\,\frac{\partial p_o}{\partial r} \quad , \tag{5b}$$

$$\alpha = \mathrm{div}\,\underset{\sim}{\zeta} = \frac{1}{r^2}\,\frac{\partial}{\partial r}(r^2\zeta) - \frac{\ell(\ell+1)}{\sigma^2 r^2}\,\chi \quad , \tag{5c}$$

$$\frac{1}{r^2}\frac{\partial}{\partial r}\left(r^2\frac{\partial\Phi'}{\partial r}\right) - \frac{\ell(\ell+1)}{r^2}\Phi' = -4\pi G\left(\zeta\frac{\partial\rho_o}{\partial r} + \alpha\rho_o\right). \quad (5d)$$

Here $\underset{\sim}{v}' \equiv i\sigma\underset{\sim}{\zeta}e^{i\sigma t}$, $\zeta \equiv \underset{\sim}{\zeta}\cdot\underset{\sim}{1}_r$, $\chi' \equiv \Phi' + p'/\rho_o$, and

$$A = \frac{1}{\rho_o}\frac{\partial\rho_o}{\partial r} - \frac{1}{\Gamma_1 P_o}\frac{\partial p_o}{\partial r} = \frac{1}{H_p}\frac{\chi_T}{\chi_\rho}(\nabla - \nabla ad), \quad (6)$$

where H_p is the pressure scale height, $\chi_T \equiv (\partial \ln p/\partial \ln T)_\rho$, and $\chi_\rho \equiv (\partial \ln p/\partial \ln \rho)_T$.

The solution of the system (5) is well-known to yield three different types of modes (cf. Cowling 1941, Ledoux and Walraven 1958, Cox 1974), termed the p-, g-, and f-modes. With the aid of Cowling's approximation ($\Phi'=0$) and Castor's approximation (cf. Cox 1974), the frequencies of the p-mode and g-mode oscillations are approximately given by

$$\sigma_p^2 \approx \left[k^2 + \frac{\ell(\ell+1)}{r^2}\right]\frac{\Gamma_1 p_o}{\rho_o}, \quad (7a)$$

$$\sigma_g^2 \approx N^2/\left[1+\frac{k^2 r^2}{\ell(\ell+1)}\right], \quad (7b)$$

while the f-mode frequencies are intermediate between these two. Here k^{-1} is the radial length scale of the oscillations, $(\Gamma_1\ po/\rho o)$ is the square of the sound speed, and $N^2 = -Ag$ is the square of the Brunt-Väisälä frequency, where g is the gravitational acceleration. Thus p-mode frequencies are ~ $(GM/R^3)^{1/2}$ ~ (10 seconds)$^{-1}$ and are insensitive to temperature, while g-mode frequencies depend upon temperature through both $(\nabla-\nabla ad)$ and $\chi_T \propto P_{ion}/P_{el} << 1$. Note that A<0 (stability against convection) implies $\sigma_g^2 > 0$: these are the dynamically stable "g^+-modes." Conversely for A>0 (convective instability) dynamically unstable "g^--modes" occur. For a given ℓ-value we follow the practice of denoting those modes with precisely n nodes in the radial eigenfunction as p_n- or $g_n^{\pm}$ - modes.

For the oscillation mode specified by $(n,\ell,m) \equiv j$, the effects of the neglected terms in equations (4) can be assessed by perturbation methods. Thus the non-adiabatic (s') terms lead to an exponential time-dependence $e^{-\kappa t}$ of the oscillations, where (cf. Cox 1974)

$$\kappa_j = - \frac{\int (\Gamma_3-1)(\delta\rho/\rho)_j^* \delta(\varepsilon-\text{div } \underset{\sim}{F}/\rho)_j \rho_o d^3r}{2\sigma_j^2 J_j} \tag{8}$$

with

$$J_j \equiv \int \underset{\sim}{\zeta}_j \cdot {}^* \underset{\sim}{\zeta}_j \rho_o \, d^3r, \tag{9}$$

and δ denotes a Lagrangian variation. In a similar manner, uniform rotation with angular velocity $\underset{\sim}{\Omega} \equiv \Omega \underset{\sim}{1}_z$ is found to lead simply to a shift in the oscillation frequency (as seen by an observer at rest) from σ_j to $\sigma_j + m\Omega(1-C_j)$, $m=-\ell, -\ell+1, ---, +\ell$, where (Ledoux 1951)

$$m\Omega C_j = \frac{i\int (\underset{\sim}{\Omega} \times \underset{\sim}{\zeta}_j) \cdot \underset{\sim}{\zeta}_j^* \rho_o d^3r}{J_j} \quad , \tag{10}$$

(See Smeyers and Denis 1971 for a self-consistent second-order calculation of rotational effects). The effects of the "thermal imbalance" terms, tidal perturbations, and convective interactions are more complex, and most of these terms have yet to be studied in the context of degenerate dwarfs. Thus we merely note here that recent discussions of these three effects can be found in the papers of Aizenman and Cox (1975), Denis (1972), and Gabriel _et al_. (1975), respectively.

4. Calculations for Specific Models

The first investigation of g-mode oscillations in white dwarfs appears to have been that of Baglin and Schatzman (1969). Motivated by the relatively long (71 second) period of DQ Her, they obtained rough estimates for the temperature dependence of the g_1 mode periods. The full fourth-order system was first solved by Harper and Rose(1970) for two $0.75M_{\odot}$ hydrogen shell-burning models. They obtained oscillation periods for the p_o, p_1, and p_2 modes belonging to $\ell=0$ and for the g_2, g_1, f, and p_o modes belonging to $\ell=2$. For the g_1 modes the periods were $\Pi_{g_1}=30.4$ seconds at $L=58L_{\odot}$ and $\Pi_{g_1} = 42.2$ seconds at $L = 377L_{\odot}$. These are comparable to the periods observed in the Z Cam stars. More recently Böhm, Ledoux, and Robe (quoted in Ledoux 1974) have obtained the periods of the

g_6 through g_1, f, and p_1 through p_4 modes of a $0.6M_\odot$, 12,000°K white dwarf model with a relatively extensive He-rich surface convection zone. The periods they obtained for the g_1 and g_2 modes, Π_{g_1} = 200 seconds, Π_{g_2} = 250 seconds are very reminiscent of the two periods observed in R548 (213 seconds and 273 seconds: Lasker and Hesser 1971).

A systematic investigation of the dependence of the non-radial oscillation periods upon the thermal structure of the unperturbed degenerate model has quite recently been carried out by Brickhill (1975). He employed Cowling's (1941) approximate second-order system of equations ($\Phi' = 0$), which are expected to be quite satisfactory for degenerate dwarfs; utilized moderately realistic static (non-evolutionary) models for the unperturbed star; and studied two groups of models: i) "white dwarfs" with masses ranging from 0.2 to 0.8 $M_\odot$ and effective temperatures mostly clustered around Te ~ 8350°K, and ii) hydrogen shell-burning models with masses of 1.0 and $1.2M_\odot$ and Te $\gtrsim$ 20,000°K ($10^{-2}L_\odot \lesssim L \lesssim 10\ L_\odot$). For the former case he has computed the g-mode periods belonging primarily to ℓ=1 and ℓ=2 (up to ℓ=4 in one case), and he has given the "excitation energy" E[$\propto$ Jj: cf. equation (9)] for the g_1 modes. His results for the g_1 and g_2 mode periods again suggest an identification of the oscillations of R548 with these modes. He also finds that the presence of the surface convection zone does not significantly affect the oscillation periods, in agreement with the conclusions of Böhm, Ledoux, and Robe. The eigenfunctions of the lowest ℓ=2 g-modes, however, are strongly concentrated toward the surface. This both reduces the "excitation energy" and enhances the possibility of excitation of the oscillations by mechanisms located in the convection zone.

Interestingly, <u>none</u> of the oscillation periods found by Brickhill are long enough to be identified with, e.g., the 750 second period of HL Tau 76. The $\ell=1$ g-modes, however, are significantly longer than the $\ell=2$ modes - for example, in a $0.8M_{\odot}$, 8340°K model, $\Pi g_1 = 236$ seconds and $\Pi g_2 = 299$ seconds for $\ell=2$, while $\Pi g_1 = 408$ seconds and $\Pi g_2 = 511$ seconds for $\ell=1$. While it is not easy to understand how the $\ell=1$ modes could be excited in an <u>isolated</u> star, it seems not improbable that they could be excited by tidal interactions in a close binary system. In this regard Fitch's (1973) interpretation of the oscillation spectrum of HL Tau 76 is particularly interesting. He finds three periodicities, at 494.22 seconds, 746.16 seconds, and ~ 3.24 hours, the last of which he suggests may be the orbital period of a faint companion. If this is correct, and if tidal forces can excite the $\ell=1$ mode, Brickhill's results suggest a larger mass than $0.8M_{\odot}$ for this star.

None of the papers discussed above have considered the problem of the damping or excitation of non-radial oscillations, however, and only one (Harper and Rose 1970) has been based upon evolutionary models (so that thermal imbalance effects can be investigated). Both of these shortcomings were alleviated in an important paper by Osaki and Hansen (1973). Their calculations were carried out for several models selected from the pre-white-dwarf evolutionary sequences of ^{56}Fe stars computed by Savedoff, Van Horn, and Vila (1969), and they calculated g_2, g_1, f, p_1, and p_2 modes belonging to $\ell=2$ as well as the radial eigenfunctions and damping integrals for these models. They showed that the g_1 mode periods for the white dwarfs obeyed the simple period-luminosity relations

$$\log \Pi_{g_1} \text{(seconds)} = 1.587 - 0.178 \log L/L_{\odot},\ M = 0.398 M_{\odot} \quad (11a)$$

$$= 1.331 - 0.171 \log L/L_{\odot},\ M = 1.0 M_{\odot}, \quad (11b)$$

which can be readily understood on the basis of equations (3) and (6), together with the relation $\chi_T \propto T$.

Perhaps the most significant aspect of their work, however, was their discussion of the damping mechanisms for non-radial oscillations (there are no driving mechanisms in the models they studied). In addition to the radiative and neutrino loss damping processes familiar from analyses of radial pulsations, they recognized the potential of non-radial oscillations for the generation of gravitational waves, and they investigated this energy loss mechanism as well. In fact, they found gravitational radiation to be the dominant damping mechanism for the p- and f- modes. For the $1.0M_{\odot}$, ^{56}Fe model with $\log L/L_{\odot} = -1.934$ (model 9N), their results are summarized in Table 2, where all units are c.g.s. except for the damping times, κ^{-1}, which are in years. Following Osaki and Hansen we have written the components of equations (8) and (9) in the forms, e.g.,

$$E \equiv \frac{1}{2}\sigma^2 \int |\underset{\sim}{\zeta}|^2 \rho_o d^3r \ , \quad L_\nu \equiv -\frac{1}{2}\int (\delta T/T)^* \delta\varepsilon_\nu \rho_o d^3r, \text{ etc.,} \qquad (12)$$

with $\zeta(r) = 1$ at the surface. It is important to note that both of these quantities are proportional to the square of the oscillation amplitude.

Table 2

Mode	Π(sec)	E	L_{ph}	L_ν	L_{GW}	κ^{-1}(ph+ν)	κ^{-1}(total)
p_2	1.009	1.0 +49	1.6 +34	1.5 +32	6.1 +39	4.0 +7	104.
p_1	1.493	2.8 +49	5.7 +33	3.6 +32	3.3 +40	3.0 +8	54.
f	2.703	1.1 +50	1.2 +33	1.5 +32	1.7 +41	4.9 +9	40.
g_1	45.98	1.3 +47	2.5 +36	1.1 +31	4.0 +25	3.3 +3	3.3 +3
g_2	62.96	3.8 +46	8.0 +36	1.9 +31	5.7 +24	3.0 +2	3.0 +2

Quite recently, Hansen, Lamb, and Van Horn (1975) have carried out this same type of calculation for the pure ^{12}C evolutionary models constructed by Lamb (1974: see also Lamb and Van Horn 1975). These models are of interest because the treatment of the physical state of the interior is believed to be the most accurate currently available for homogeneous stars in these phases of evolution. A fully temperature-dependent treatment of the Coulomb interactions is used, which includes the effects of crystallization and Debye cooling, and self-consistent carbon convective envelope models are employed. The results for a model with $\log L/L_{\odot} = -1.928$ are briefly summarized in Table 3 for comparison with the calculations of Osaki and Hansen.

Table 3

Mode	Π	$\kappa^{-1}(ph+\nu)$	$\kappa^{-1}(tot)$
p_2	1.496	4.7 +7	530.
p_1	2.217	2.7 +8	190.
f	4.204	6.8 +9	120.
g_1	116.60	1.2 +3	1.2 +3
g_2	122.82	3.6 +1	3.6 +1

The dominance of gravitational radiation damping for the f- and p- modes is again clear, and the damping rates are not grossly different in spite of the rather larger oscillation periods of the Lamb models. In addition, we have investigated the effects of thermal imbalance, following Aizenman and Cox (1975), and have found them to be completely negligible for these modes, at typical white dwarf luminosities. We have also found a hint of secular instability due to strong driving in the partial ionization zone in the g_1^+ and g_2^+ modes of one model with log $L/L_\odot$ = -0.167 (log Te = 4.765). Whether this result is real, however, remains to be established. Evidently further studies of the secular stability of models with realistic surface convection zones and various masses and compositions are sorely needed.

5. Summary and Conclusions

The dependence of the non-radial mode oscillation periods upon the thermal properties of the unperturbed degenerate star models are beginning to be understood. The magnitudes of the $\ell=2$ g-mode periods in 1.0 to $1.2M_\odot$ hydrogen shell-burning stars are of the correct magnitude to account for the periodicities in the Z Cam stars, and rotation splitting of these modes, as suggested by Warner and Robinson (1972) to account for the observed fine structure, seems plausible. The case of the variable white dwarfs is in less satisfactory shape, however. Although the $\sim 10^2$ second periodicities seem understandable as low-order $\ell=2$ g-modes, as in the case of R548, the origin of the $\sim 10^3$ second oscillations is still mysterious. Tidally-driven $\ell=1$ g-modes seem promising, but much work remains to be done.

Probably the most important current theoretical problems concerning non-radial oscillations of degenerate dwarfs are the following.

i) The question of the existence of g^--modes. These are expected to be present on the basis of calculations for other types of stellar models, but *none* have yet been discovered in any of the degenerate dwarf models. This may simply be due to the difficulty of locating these modes in models with convection zones as thin as those in white dwarfs, but it is important that this be resolved, since the presence of g^--modes may have important consequences for the excitation mechanisms of non-radial oscillations.

ii) The nature of the excitation mechanisms for non-radial modes. The strong gravitational radiation damping of the p- and f- modes in white dwarfs makes these modes difficult to excite. In addition, the fact that the eigenfunctions of the low-order g-modes are largest near the surface (where the potential excitation mechanisms are concentrated in white dwarfs) enhances the possibility of excitation of these modes. Aside from the single tentative case of secular instability (due to the usual κ- and γ-mechanisms: cf. Cox and Giuli 1969) recently found by Hansen, Lamb, and Van Horn, however, there have been no investigations of de-stabilitzing influences. Such studies are sorely needed.

In addition, the possibility of nuclear excitation of the oscillations needs to be investigated both for the shell-burning stars and for the DA (hydrogen spectra) white dwarfs, which may have relatively deep hydrogen surface layers. For white dwarfs of other spectral types, however, nuclear excitation due to hydrogen-burning at the base of the convection zone appears unlikely to be significant. If we set Osaki and Hansen's (amplitude-dependent) L_{ph}, which is the dominant damping mechanism for g-modes, equal to the product $\varepsilon_b \Delta M_c$ from Table 1, which provides a generous upper limit to the maximum thermonuclear luminosity, we find relative oscillation amplitudes $\lesssim 10^{-5}$.

iii) The possibility of tidal excitation in a close binary system. The fact that the $\ell=1$ modes, which probably can be excited only in a binary system, have substantially longer periods than the modes belonging to higher ℓ-values indicates the potential interest of these modes. The fact that many of the variable white dwarfs are now suspected or known close binaries makes it important to investigate this problem in more detail. If $\ell=1$ modes are the explanation of the $\sim 10^3$ second periodicities, observational tests of this mechanism may also be possible.

iv) Mode-coupling. If dynamically unstable $\bar{g}$-modes do exist in white dwarfs, mode coupling may permit the excitation of otherwise stable modes, as recently suggested by Osaki (1974) for the β Ceph stars. Coupling with convective motions (if these are different from $\bar{g}$-modes) also seems likely to be important because of the near commensurability of the g^+ mode periods and convective time-scales. These processes may also have a bearing on the low-frequency "flickering" found in some of the variable white dwarfs (e.g., CD - 42° 14462: Hesser, Lasker, and Osmer 1974).

v) Finally, the fact that crystallization is expected to occur in some of the more massive and fainter stars suggests the interesting possibility of whole new classes of stellar oscillations and instabilities. The solid core can sustain shear modes, and the solid-liquid core interface may cause significant changes in the oscillation spectrum as the core grows in the cooling star. The possibility of erratic excitation of oscillations triggered by "star quakes" in the cooling and contracting solid core also immediately suggests itself.

Acknowledgements

I am grateful to Drs. C.J. Hansen and M.L. Aizenman for many discussions of the theoretical problems of non-radial oscillations and thermal imbalance in degenerate stars; to Drs. J.G. Duthie, R.A. Berg and S. Starrfield for discussions about the observational data; and to Dr. A.J. Brickhill for providing me with a copy of his manuscript in advance of publication. This work has been supported in part by the National Science Foundation under grant MPS 74-13257.

References

Aizenman, M.L., and Cox, J.P.: 1975, Astrophys. J. 195, 175.

Baglin, A. and Schatzman, E.: 1969, in Low Luminosity Stars (ed. by S.S. Kumar), Gordon and Breach, New York, p. 385.

Baglin, A. and Vauclair, G.: 1973, Astron. Astrophys. 27, 307.

Böhm, K.-H.: 1968, Astrophys. Space Sci. 2, 375.

Böhm, K.-H.: 1969, in Low Luminosity Stars (ed. by S.S. Kumar), Gordon and Breach, New York, p. 357.

Böhm, K.-H.: 1970, Astrophys. J. 162, 919.

Böhm, K.-H. and Cassinelli, J.P.: 1971a, in White Dwarfs (ed. by W.J. Luyten), D. Reidel Publ. Co., Dordrecht, p. 130.

Böhm, K.-H. and Cassinelli, J.P.: 1971b, Astron. Astrophys. 12, 21.

Böhm, K.-H. and Grenfell, T.C.: 1972, Astron. Astrophys. 28, 79.

Brickhill, A.J.: 1975, preprint.

Chandrasekhar, S.: 1935, Monthly Notices Roy. Astron. Soc. 95, 207.

Chandrasekhar, S.: 1957, An Introduction to the Study of Stellar Structure, Dover Publ., Inc., New York.

Chanmugam, G.: 1972, Nature Phys. Sci. 236, 83.

Clayton, D.D.: 1968, Principles of Stellar Evolution and Nucleosynthesis, McGraw-Hill Book Co., New York.

Cowling, T.G.: 1941, Monthly Notices Roy. Astron. Soc. 101, 368.

Cox, J.P.: 1974, unpublished lecture notes, University of Colorado.

Cox, J.P. and Giuli, R.T.: 1968, Principles of Stellar Structure, Gordon and Breach, New York, Chapter 27.

D'Antona, F. and Mazzitelli, I.: 1974, Astrophys. Space Sci. 27, 137.

D'Antona, F. and Mazzitelli, I.: 1975, Astron. Astrophys. to be published.

Denis, J.: 1972, Astron. Astrophys. 20, 151.

Faulkner, J. and Gribbin, J.R.: 1968, Nature, 218, 734.

Fitch, W.S.: 1973, Astrophys. J. 181, L95.

Fontaine, G.:1973, Ph.D. thesis, University of Rochester.

Fontaine, G., Van Horn, H.M., Böhm, K.-H., and Grenfell, T.C.: 1974, Astrophys. J. 193, 205.

Gabriel, M., Scuflaire, R., Noels, A., and Boury, A.: 1975, Astron. Astrophys. 40, 33.

Hamada, T. and Salpeter, E.E.: 1961, Astrophys. J. 134, 683.

Hansen, C.J., Lamb, D.Q., and Van Horn, H.M.: 1975, to be published.

Hansen, J.P.: 1973, Phys. Rev. 8A, 3096.

Harper, R.V.R. and Rose, W.K.: 1970, Astrophys. J. 162, 963.

Hesser, J.E., Lasker, B.M., and Osmer, P.S.: 1974, Astrophys. J. 189, 315.

Lamb, D.Q.: 1974, Ph.D. thesis, University of Rochester.

Lamb, D.Q. and Van Horn, H.M.: 1975, Astrophys. J., in press.

Ledoux, P.: 1951, Astrophys. J. 114, 373.

Ledoux, P.: 1974, in Stellar Stability and Evolution (ed. by P. Ledoux, A. Noels, and A.W. Rodgers) D. Reidel Publ. Co., Dordrecht, p. 135.

Ledoux, P. and Walraven, Th.: 1958, Handbuch der Physik 51, Springer-Verlag, Berlin, p. 353.

Mestel, L.: 1952, Monthly Notices Roy. Astron. Soc. 112, 583.

Osaki, Y.: 1974, Astrophys. J. 189, 469.

Osaki, Y. and Hansen, C.J.: 1973, Astrophys. J. 185, 277.

Richer, H.B. and Ulrych, T.J.: 1974, Astrophys. J. 192, 719.

Savedoff, M.P., Van Horn, H.M., and Vila, S.C.: 1969, Astrophys. J. 155, 221.

Schwarzschild, M.: 1958, Structure and Evolution of the Stars, Princeton University Press, Princeton.

Smeyers, P. and Denis, J.: 1971, Astron. Astrophys. 14, 311.

Van Horn, H.M.: 1971, in White Dwarfs (ed. by W.J. Luyten), D. Reidel Publ. Co., Dordrecht, p. 97.

Warner, B.: 1973, Monthly Notices Roy. Astron. Soc. 162, 189.

Warner, B.: 1974, Monthly Notices Roy. Astron. Soc. 167, 61P.

Warner, B.: 1975, this Colloquium.

Warner, B. and Robinson, E.L.: 1972, Nature Phys. Sci. 239, 2.

Wheeler, J.C., Hansen, C.J., and Cox, J.P.: 1968, Astrophys. Letters, 2, 253.

Discussion to the paper of VAN HORN

BATH: The periodicities observed in dwarf novae exhibit large changes ($\sim$10%) in period on very short timescales ($\sim 10^3 - 10^4$ sec). If interpreted (as you have implied) as due to varying g-mode periods associated with changes of luminosity of equilibrium models, then the structural changes within the dwarf are enormous. They require internal energy changes of $\sim 10^{45}$ erg - much larger than the outburst energy itself. Brickhill agrees with this in the paper you refer to.

If one assumes that runaway nuclear burning occurring in the surface layers can change the period in the way observed, that is a different problem. But in that case, none of the work on g-mode pulsations presently published is relevant to your argument.

VAN HORN: I agree that if you consider equilibrium white dwarf models, then the observed period changes do indeed imply unacceptably large changes in the internal energy. However, in the case of degenerate, hydrogen-shell-burning stars, it does not seem impossible to me that the rapidly changing thermal structure of the envelope - especially the distension of the hydrogen-rich surface layers - may be able to produce the observed period changes. Brickhill's calculations are indicative of this, and they are certainly relevant to this point, but calculations must clearly be extended to rapidly evolving, hydrogen-shell-burning, degenerate stars before these indications can be regarded as being either confirmed <u>or</u> rejected.

BATH: Since I shall not have time to discuss the periodicity problem of dwarf novae in my talk, it should be pointed out here that the periods so far observed all fall in the range 16-35 sec. The period with which matter orbits the dwarf surface at the inner boundary of the accretion disc is in the range 14-55 sec for white dwarf masses $\sim$1.4-0.1 $M_\odot$. Clearly, the regular eclipse by the dwarf of inhomogeneities (or luminosity fluctuations) in the central regions of the disc could account for the observed periodicities. The changing periods then only imply small changes in the orbital radius of the inhomogeneity. Accurate determination of the lifetime of such a disc fluctuation is obviously a difficult problem.

VAN HORN: I agree that the <u>periodicity</u> of matter at the inner edge of an accretion disk around a white dwarf is of the correct order of magnitude to account for the observed periods. Whether the systematic properties of the observed periodicities can be satisfactorily explained by this hypothesis, however, is another matter. To mention only two points that bother me: (1) It is not obvious to me that something as fragile as I expect an accretion disk to be, can sustain fluctuations long enough to account for the observed coherency of the oscillations, and (2), I would expect the fluctuations to be carried inward through the disk, and this will cause the period to <u>decrease</u> rather than increase as observed.

Multiple Periodic Variable Stars
IAU Colloquium No.29, Budapest, 1975

THE RS CVn BINARIES AND BINARIES WITH SIMILAR PROPERTIES

Douglas S. Hall

Dyer Observatory, Vanderbilt University

I. Introduction

This paper will review reasonably thoroughly and comprehensively the many observational properties of the remarkable RS CVn-type binaries. In addition this paper will show that many of these same properties are observed in other types of binary systems.

One difficulty in discussing the RS CVn binaries has been that the characteristics required for membership in the group have never been agreed upon. Therefore this matter is discussed in Section II and a working definition proposed: binaries with orbital periods between 1 day and 2 weeks, with the hotter component F-G V-IV, and with strong H and K emission seen in the spectrum outside eclipse. The remainder of Section II will review the many observed properties and physical characteristics of the RS CVn binaries. See Table 1. Periodicities and interrelationships among properties will be described. It is easy to see how much has been learned in the last seven years. The last time RS CVn was discussed in Budapest, the title of the Colloquium was "Non-Periodic Phenomena in Variable Stars". Now RS CVn is being discussed again in Budapest, but the title of the Colloquium is "Multiply Periodic Phenomena in Variable Stars".

Section III will treat related types of binaries. First are binaries with orbital periods longer than 2 weeks in which one component is of spectral class G-K IV-II and displays strong H and K emission. See Table 2. This group will be referred to as the long-period group. Second are the non-contact binaries with periods less than 1 day in which the hotter component is F-G V-IV and H and K emission is displayed in one or both components. See Table 3. This group will be referred to as the short-period group. Third are binaries consisting of dwarf K or dwarf M stars which display H and K emission. This group will be referred to as the flare star group. Fourth is the unique binary V471 Tau = BD +16°516. And fifth are W UMa binaries for which there is photometric evidence of uneven surface brightness on one component. This group will be referred to as the W UMa group.

I have tried to make my list of RS CVn binaries complete and the data in Table 1 up-to-date as of mid-1975. The same cannot be said for the other groups and the data in their respective tables. Locating the available information for the RS CVn binaries was difficult because much of it can be found only in observatory reports, unpublished theses, and private communications. Tables 1, 2, and 3 contain only one value for each quantity, but a reference to its source is given. The value chosen is the one I considered best, but the reference in most cases enables the reader to reconsider my decision.

In cases where more than one interpretation is possible, more time is spent discussing the one I consider most reasonable. In general, however, reference is made to alternate interpretations which merit consideration.

II. THE RS CVn BINARIES

A. Introduction

A complete, up-to-date list of proposed condidates for this group has not been published. The latest is that of Popper (1970). The unpublished thesis of Oliver (1974a), besides presenting many new results and important conclusions and a more up-to-date list of members, contains a good review of the older literature and provides the key to many unpublished results. Thus it was useful as a starting point for this review.

B. Historical Development of the Definition of the Group

From an historical point of view the definition of the group has proceeded as follows. Struve (1946) was probably the first to call attention to the existence of the group. He discussed 5 binaries with H and K (but not Balmer) emission lines which were visible outside eclipse, arose from a late G or early K component, seemed to reflect the orbital motion of that component, and were much stronger than would be expected in typical single stars of the same spectral class. Hiltner (1947) listed 13 binaries which he felt belonged to such a group. Gratton (1950) discussed a group of 19 late-type H and K emission binaries which included Hiltner's 13. The other 6 were generally longer in orbital period, more luminous, and not eclipsing variables. In a study of the absolute dimensions of eclipsing binaries Plavec and Grygar (1965) concluded that one group stood apart from the rest of the Algol-type binaries. Both components were late-type, the mass ratio was around unity, and the binary seemed to be detached. Included were RS CVn, WW Dra, Z Her, AR Lac, and SZ Psc. In his discussion of mass and radius determinations in eclipsing binaries Popper (1970, Table 2) presents a list of 22 eclipsing binaries which represents the most comprehensive list of possible candidates for the group published to date. Here he mentions the H and K emission, the curious fact that the component responsible for the emission usually is around KO IV, the tendency for the mass ratio to be near unity, and the intrinsic variability in one or both components. Preliminary reports by Oliver (1971, 1973) drew further attention to the existence of this group and provided a more complete list of the remarkable properties. The unpublished thesis of Montle (1973) considered adding RT And, KO Aql, RZ Cnc, ER Vul, and the spectroscopic binaries HD 21242 = UX Ari and HD 209813 = HK Lac to the list of Popper. Oliver (1974a) bases his comprehensive study on the 22 binaries in Popper's list plus RT CrB and GK Hya.

C. A Proposed Working Definition for an RS CVn Binary

Oliver (1974a, p. 262) was the first to propose formally a set of observational characteristics to define the group. But this approach is not entirely satisfactory because, as he pointed out, some of his defining characteristics need not be present in all binaries. Therefore I want to consider this question of definition anew.

I think it is necessary to consider orbital period. One reason is the similarity between the long-period group (Herbst 1973) and the RS CVn group, which was first pointed out by Gratton (1950). Despite the obvious similarities, it is not yet known how the two groups are related to each other. Therefore, in case they are fundamentally different, it is safer to make a distinction; but, in case they are fundamentally related, we should not forget the similarities. It turns out convenient (and perhaps significant) to take advantage of the absence of suspected candidates in either group with orbital periods in the range 11 to 17 days. In addition quite a few RS CVn-type properties occur in several short-period binaries (RT And, SV Cam, CG Cyg, and ER Vul) not generally considered members of the RS CVn group. This made me realize that the RS CVn group might need definition at the short-period end also. Since Oliver (1974a, p. 262) expressed the same concern for different reasons, I was encouraged to follow his tentative suggestion that systems with orbital periods less than 1 day be excluded from the RS CVn group. This takes advantage of the absence of suspected candidates in the range $0^{d}.9$ to $1^{d}.9$. Thus I propose that the RS CVn group contain binaries only in the orbital period range 1 day to 2 weeks.

The next step is to compile a list of RS CVn binaries, starting with the 24 systems considered by Oliver (1974a, Table 1). The orbital period criterion eliminates RZ Eri, WY Cnc, and UV Psc. Then there are 3 additional binaries which should be added: UX Com (Popper 1974), HD 21242 = UX Ari (Evans and Hall 1974), and HD 118216 (Conti 1967). The last two are not eclipsing variables. Now we have a list of 24 binaries which I propose as members of the RS CVn group. These are listed in Table 1 along with a summary of their various observational and physical properties.

At this point I think the best strategy is to examine these 24 binaries and note which characteristics are exhibited by <u>all</u> 24. This should give us a reasonably restrictive working definition for an RS CVn binary which should be useful in the search for additional members. Thus the proposed definition is as follows: The orbital period is in the range 1 day to 2 weeks, the hotter component is of spectral type F or G and luminosity class V or IV, and strong H and K emission is seen outside eclipse. The term "strong" here means stronger than the normal Wilson-Bappu emission observed in the H and K reversal in single stars. It is interesting to note that this definition automatically excludes the three other related groups: the flare star group, V471 Tau, and the W UMa group. Needless to say this definition can, step by step, be made more restrictive if accumulation of better and more complete data shows us that other properties are shared by all or virtually all systems.

Somewhat confusing is the fact that not everyone agrees what the name of this group should be. Most people name the group after RS CVn as the prototype, but some refer to AR Lac as the prototype. A few are careful to stress the fact that, until now, the group has not officially been defined. Consequently they have hesitated to name a prototype and instead have referred to the group with a sentence or a phrase which summarizes the characteristics. Let me propose RS CVn as the prototype of the group defined in this paper, since it is at present generally considered as such and is an appropriate choice.

D. Observed Characteristics and Physical Properties

This subsection discusses the large collection of fascinating observational characteristics and physical properties exhibited by the 24 RS CVn binaries in Table 1.

1. Light Curve Variations

To me the multiply periodic variations in the light curve are the most fascinating aspect of the RS CVn binaries. Before the middle of the 1960's it appeared that the light curves just varied irregularly with a variety of time scales. A breakthrough was made by the astronomers at Catania (Chisari and Lacona 1965; Catalano and Rodono 1967, 1969) when they observed a persistent nearly-sinusoidal wave-like distortion in the light curve of RS CVn outside eclipse ($\approx 0\overset{m}{.}2$ from maximum to minimum) which migrated slowly (one cycle every $\approx$ 10 years) towards decreasing orbital phase. They also showed that this wave migration was correlated with a variable depth of primary minimum and a variable displacement of secondary minimum.

One consequence of the wave is to render the two maxima unequal in brightness. This particular phenomenon in eclipsing binaries was first investigated systematically by Mergentaler (1950) and by O'Connell (1951). The migration of the wave causes the sense of the inequality to change with time. This also was observed in the past.

Oliver (1971, 1973, 1974a) made the important finding that these persistent waves and their migration towards decreasing orbital phase seem to be a common property of the RS CVn binaries in general. A look at Table 1 shows that 15 are now known or suspected to have waves, of which 8 are known or suspected of migrating. The migration periods range from about 5 years to about 75 years.

Hall (1972) analyzed in some detail the available photometry of RS CVn and showed that a simple model could account for a large number of observed properties, periodicities, and interrelationships. He considered the H and K emission as indication of chromospheric activity and made use of analogies with sunspot activity in our sun. In his model a region of large-scale spot activity darkens one side of the cooler star within 30° of its equator. This darkening produces the wave, explains its red color, and accounts for the anomalously shallow depth of secondary minimum. A compromise between differential rotation (like that observed in our sun) and synchronous rotation (like that observed in most binaries with rather small separations) produces the migration of the wave towards decreasing orbital phase. The migrating wave naturally explains the variable depth of primary minimum and the variable displacement of secondary eclipse. A 23.5-year "sunspot cycle" operating in the cooler star accounts for the variable amplitude of the wave, which ranges between $0^{m}_{.}2$ and $0^{m}_{.}05$. This cycle also causes the spots to drift periodically in latitude, as is seen in the so-called butterfly diagram of sunspot activity, and thereby can account for the non-uniformity of the migration rate, the period of which varies between 8 and 12 years every 23.5 years.

Table 1 shows that there are other examples of variable amplitudes and other examples where the migration rate is not constant. Variable depth of primary minimum and displacement of secondary eclipse are not considered in Table 1 among the observable properties because they can be considered consequences of the migrating wave.

Table 1 lists binaries with irregular light curve changes. These are changes not accounted for by a migrating wave, a variable wave amplitude, a variable migration period, and associated changes in the depth of primary minimum and displacement of secondary eclipse. Examples would be changes in the overall level of brightness from one season to the next, a disruption in the shape or phase of the wave, or short-time-scale flaring. Such irregular changes have been suspected in certain binaries but until recently there was some reason to doubt their validity. To date the only published work illustrating such a change is the UBV photometry of UX Ari (Evans and Hall 1974), which indicates that the overall light level decreased by over $0^{m}.1$ sometime between late 1972 and late 1974. Irregular changes in AR Lac have been reported by both Wood (1946) and Kron (1947) but unfortunately they both used HD 209813 = HK Lac as a comparison star, which Blanco and Catalano (1970) showed to be a variable. It is cruel irony that HK Lac appears in Table 2 of this same review paper as a member of the long-period group, the first one in which a migrating wave was observed. The HK in the variable star designation provides additional irony. And the final irony is that more recent photoelectric photometry based on different comparison stars (Babaev 1971, Chambliss 1975) shows that AR Lac does exhibit irregular fluctuations in brightness afterall.

2. Spectroscopic Characteristics

It follows from the definition proposed in this paper that all 24 binaries in Table 1 display strong H and K emission lines outside eclipse. In virtually every one the emission can be attributed to the cooler component. For the newly added member UX Com I do not have enough information to decide; Andersen and Popper (1975) say that the two components in TY Pyx have the same spectral type; RV Lib appears to be the one exception to the rule. The earlier component also displays H and K emission in 4 of the binaries; Popper (1970) thinks this tends to happen when the hotter component is cool enough.

In addition to H and K emission, several of the binaries show Hα in emission. RS CVn also shows Hβ emission (Naftilan 1975). There is some tendency for this emission to arise from the cooler component. Weiler (1975a) observed 6 RS CVn binaries, selected only on the basis of apparent brightness, and detected Hα emission in all 6.

An important point which has been stressed most recently by Popper (1970) and Oliver (1974a) is that the H and K emission lines give velocities which always agree well with the absorption line velocities arising from the same component. Another important point (Popper 1970, p. 53) is that the width of the emission line corresponds approximately to that of a photosphere rotating synchronously with the orbit. A third point is that the radial velocity curves from both components are remarkably well-defined, free of distortion and scatter, and usually indicative of circular orbits. From these three points one can safely conclude that the H and K emission is fundamentally different in its origin from the Balmer emission in Algol-type binaries, which arises from faster-than-synchronously rotating circumstellar rings around the hotter star or from gas streams connecting the two stars.

Several spectroscopic investigators have studied the behavior of H and K emission with orbital phase. Oliver (1974a, p. 251) summarizes these by saying that the H and K emission from the cooler component seems sometimes to be concentrated on the stellar disk so as to be partially or totally obscured during secondary minimum. Hiltner (1947) claimed that there was a tendency for the H and K emission to originate from the opposite ends of the star in question, the ends elongated by tidal distortion. This interpretation has been repeated by others since then, even though the polarimetric observations of AR Lac by Struve (1948) failed to find the degree of polarization which would be expected in such an interpretation.

Most recently Weiler (1975b) made an important finding. In the three binaries UX Ari, RS CVn, and Z Her he finds that the intensity of both H and K emission and Hα emission appeared to be correlated with orbital phase, which is equivalent to correlation with wave orientation since his observations were made within an interval of about one year. In the case of UX Ari and RS CVn there were sufficient photometric data available to fix the orientation of the migrating wave in 1974, the year of his observing program. In both cases he found that maximum H and K emission and maximum Hα emission coincided very closely with phases when the fainter hemisphere faced the earth. Since Weiler finds emission strength correlated with wave orientation and not phase of conjunction or quadrature, his observations argue against the Hiltner interpretation and provide support for the model of a migrating spotted region.

Oliver (1974a, p. 251) concluded also that there were variations in the absorption line strengths but that they were not well correlated with light curve variations.

Abundance anomalies have been reported for some of the RS CVn binaries by Miner (1966), Hall (1967), and Naftilan (1975). Somewhat surprisingly these three find heavy elements apparently <u>under</u>abundant, but the significance of this finding is not clear. I think this underabundance should not be regarded as real until the effect of possible veiling in the ultraviolet and violet has been allowed for. To my knowledge lithium has been searched for in only one RS CVn binary. Conti (1967) looked for lithium in his spectra of HD 118216 and was unable to detect it in the F5 component.

3. Physical Parameters

Reliable masses are relatively easy to determine for the RS CVn binaries (Popper 1967) because their spectra tend to be two-lined and because, as mentioned above, radial velocities derived from the absorption <u>and</u> the emission lines are not distorted by the influence of gas streams and other masses of circumstellar material as they are in the Algol-type binaries. The most recent and complete list of reliable masses is given by Oliver (1974a, Table 47). These are given in Table 1 along with a few from other sources. I have included none which were not derived from two-lined spectrograms.

The data in Table 1 show that for all but two binaries the mass ratio tends to be very close to unity. With AD Cap and RT Lac excluded, the average mass ratio is 1.03 ± 0.03 (rms). The rms deviation of a single value from this mean is only ± 0.1, part of which must be observational uncertainty. It is perhaps significant (but puzzling) that the only two binaries known to have mass ratios markedly different from unity (AD Cap and RT Lac) are among the 4 (out of 24) systems in which <u>both</u> components display H and K emission.

Of the RS CVn binaries in Table 1 for which both masses and radii are known, all are detached. This important characteristic was first noticed by Plavec and Grygar (1965) and most recently demonstrated by Oliver (1974a, Figure 64). The ratio of the radius of each star to that of its respective Roche lobe averages about 60% for the cooler star and about 35% for the hotter, so these RS CVn binaries are detached by no small margin.

The total mass of the RS CVn systems in Table 1 ranges between 1.75 $M_\odot$ and 2.98 $M_\odot$ if we consider only those near unit mass ratio. The lower limit becomes a bit smaller, around 1.6 $M_\odot$, if we include AD Cap.

In cases where both components have been classified, the spectral type of the hotter star ranges between F4 and G9 and the luminosity class ranges between V and IV. The tendency for the hotter star to lie slightly above the main sequence appears to be significant. As pointed out by Popper (1967, 1970), Oliver (1974a), and Andersen and Popper (1975), there is a marked tendency for the cooler component to be very near spectral class K0 IV, the only exception known to date being TY Pyx.

Distances for most of these systems were determined by Montle (1973). These are included in Table 1 because distances are not generally available from other sources.

4. Period Variations

One of the most remarkable properties is the tendency for the orbital periods to be variable. These involve both increases and decreases, are around $\Delta P/P = 10^{-4}$ or 10^{-5}, and occur on time scales of years or tens of years. O-C deviations from the best linear ephemeris can amount to as much as $\pm\ 0\overset{d}{.}25$ (Arnold, Hall, and Montle 1973). In some cases a linear ephemeris taken arbitrarily from a catalogue can, after only ~ 10 years elapsed time, lead to phases which are in error by one fourth of the orbital cycle. Adequate data are not available to tell us how many systems have variable periods, but about a third are known to so far. I suspect virtually all do.

It should be clear that quite different mechanisms are likely to be responsible for period changes in RS CVn binaries and those in Algol-type binaries even though both experience alternate period changes of comparable size and on comparable time scales. For Algol-type binaries the most likely mechanism seems to be loss of mass on a dynamical time scale from a convective star which fills its Roche lobe (Hall 1975a). In RS CVn binaries both components are smaller than their Roche lobes by $\sim 1\ R_{\odot}$ or more, so this Algol mechanism definitely cannot work.

It should also be very clear by now that these period changes cannot be explained via apsidal motion or orbital motion around a third body, simply because the observed changes are not strictly periodic, let alone sinusoidal.

In my opinion the most useful clue to understanding these period changes is the interesting correlation observed between period changes and wave migration (Hall 1975b). In the 4 RS CVn binaries RS CVn, SS Cam, AR Lac, and RT Lac and the 2 related binaries CG Cyg and V471 Tau, attempts have been made to demonstrate that period decreases occur when the minimum of the wave is around orbital phase $0\overset{p}{.}25$ and period increases occur when the minimum of the wave is around $0\overset{p}{.}75$. Catalano and Rodono (1974) pointed out that the migrating wave can distort the shape of primary minimum in a way which will generate spurious O-C variations which are in the sense of the above correlation. Hall (1975b) showed that this influence is not large enough to account for the observed O-C variation in RS CVn, CG Cyg, or SS Cam. In fact in SS Cam the observed O-C variation is $\sim$25 times larger than the maximum possible shift due to the distortion of primary eclipse. Nevertheless, Catalano and Rodono were right to emphasize that period changes in RS CVn binaries should be investigated only after effects of light curve distortion have been carefully discussed.

Given such a correlation, it is most obvious to explore mechanisms involving mass loss from either the fainter or the brighter hemisphere of the cooler star, which produces the wave. Arnold and Hall (1973) considered high-velocity impulse-type mass ejection from the brighter hemisphere. Catalano and Rodono (1974) objected to the idea of having mass lost from the brighter hemisphere, whereas one would expect flare-type mass ejection from the darker hemisphere, which is supposed to be experiencing the spot activity. Another objection is to the large amount of mass loss required: 10^{-6} $M_{\odot}$/year. But there are many other possible mechanisms to explore besides the simple impulse-type mass ejection mechanism. For example the mass might corotate with the system out to some Alfven radius. This could counter the two objections. First, using the Alfven radius as the effective moment arm instead of the orbital axis of the mass-losing star lets the observed period changes be produced with less mass loss per year. Second, since trajectories of particles ejected from flares are curved as they go from the photospheric limb to the Alfven radius, it is possible that particles leaving the binary system in the vicinity of that part of the Alfven radius lying above the brighter hemisphere of the cooler star actually were ejected from the darker hemisphere, as one would expect. In any case, this interesting problem of period changes demands more attention by observers and theoreticians alike.

Ulrich (1976) argues that an enhanced solar-wind-type mass loss of around 10^{-9} $M_{\odot}$/year is reasonable to expect from the KO IV component. One should explore the possibility that mass loss of this size is capable somehow of producing the observed period changes.

In two of the binaries, AR Lac and RT Lac, there is photometric evidence of an envelope of material surrounding one of the two components. Catalano (1973) reported observations which showed a $\approx 0^{m}.05$ depression in the light curve of AR Lac just before first contact of primary eclipse and just after fourth contact. I would interpret this as a result of occultation of the hotter star by an envelope or shell of material surrounding the cooler star, perhaps filling its Roche lobe. Hall and Haslag (1976) found the same effect in the light curve of RT Lac. It seems that an increase in the density of this shell around the beginning of World War I was responsible for the remarkable fact that the light curve of RT Lac changed from Algol-type ($A_2 \approx$ -0.03) to β Lyrae-type ($A_2 \approx$ - 0.09) at that time. Such envelopes could be an additional observable consequence of the mass outflow hypothesized to explain the period changes observed in RS CVn binaries.

5. Ultraviolet, Infrared, and Radio Observations

The most recent and complete discussion of ultraviolet excess in RS CVn binaries is that of Oliver (1974a). He says that frequently there is an indication of ultraviolet excess in the U-B color of the cooler star and that sometimes an excess is seen also in the hotter star. Photometry on a standardized system such as UBV is not available for most of the RS CVn binaries, so it is quite possible and even likely that ultraviolet excess occurs in virtually all. Using chromospheric activity as the point of comparison, Hall (1972) suggested that this excess might be analagous to that found in the T Tau variables.

Infrared excess also seems to be characteristic of most RS CVn binaries. Atkins and Hall (1972) found an infrared excess in 5 of 6 RS CVn systems for which they obtained sufficient JHKL photometry to make the decision. These 6 systems were selected only on the basis of apparent magnitude and availability in the sky, so it seems that a large fraction of the RS CVn binaries have infrared excesses. In all cases they found the excess is about $0^{m}.5$ in J, H, K, and L if it is attributed entirely to the cooler star. Since then Milone (1976) has detected an infrared excess in 2 of these 5 (RS CVn and AR Lac) and in another (RT Lac). Needless to say, infrared excess is an expectation of the model in which an appreciable area of the cooler star is affected by large-scale spot activity. Atkins and Hall pointed out that the effective temperature derived from the energy distribution between ultraviolet and infrared is appreciably cooler than that derived from the $(B-V)_0$ or the observed spectral type alone; in fact their lower effective temperature places a KO IV star on the Hayashi track for a star of $\approx 1\ M_\odot$. Milone (1976) considers circumstellar material as a possible source of the infrared excess. This interpretation should not be ignored since, as mentioned above, there are photometrically detectable envelopes in at least two systems (RT Lac and AR Lac).

One of the most remarkable recent discoveries pertaining to the RS CVn binaries is the detection of strong radio emission from several members: AR Lac, UX Ari, and RT Lac. It is remarkable that a star as faint as RT Lac ($V = 10\overset{m}{.}2$ at maximum) and relatively distant (205 parsecs) emits detectable radio emission. Gibson and Hjellming (1974) are thinking that the ultimate energy source for the radio emission in these and in the Algol-type binaries is gravitational infall of matter transferred from one component to the other. One should, however, explore the possibility that some other mechanism is responsible in the RS CVn binaries; we know that radio emission in the same frequency range is associated with sunspots in our sun and that radio emission coincident with optical flares is observed even in single stars. A crucial test would be to see if radio emission, like Hα emission and H and K emission, is correlated with phases when the fainter hemisphere of the cooler star is facing the earth.

6. Space Density, Galactic Distribution, and Kinematical Properties

The space density of the RS CVn binaries is very large. Using the distances of Montle (1973) I find a space density of 1.0 X 10^{-6} systems/pc^3. Dworak (1973) has listed 40 eclipsing binaries probably within 100 pc of the sun, of which 7 are among those listed in Table 1 of this review. This implies a space density of 1.7 X 10^{-6} systems/pc^3. The smallness of the sample in each case produces some statistical uncertainty, but not much: ± 20% for Montle and ± 35 % for Dworak. Both of these estimates are lower limits since the non-eclipsing RS CVn binaries, of which there should be many, are not being counted. Such a space density is remarkably high; the W UMa binaries, at one time thought to be the most plentiful type of binary system, have a space density of only about 10^{-6} systems/pc^3 (Kraft 1967) even when non-eclipsing systems <u>are</u> counted.

Montle determined the mean height above and below the galactic plane to be $\langle z \rangle$ = 109 ± 20 pc for his entire sample of 29 systems or $\langle z \rangle$ = 112 ± 20 pc for a restricted sample which excluded the three doubtful cases RT And, KO Aql, and HK Lac. These determinations involved careful allowance for incompleteness effects. He constructed a calibration of $\langle z \rangle$ versus age by using data on several different types of galactic objects available from a variety of published sources. The accidental uncertainty in this calibration was about ± 5 pc at the part of the curve he used. Entering the observed $\langle z \rangle$ into the calibration yielded an age of 2 X 10^8 years, and considering the uncertainties of both indicated an age in the range 1 - 4 X 10^8 years.

Montle also determined the dispersion of their velocity component perpendicular to the galactic plane to be $\langle Z^2 \rangle^{1/2}$ = 10.0 ± 2.5 km/sec. This value is the intrinsic dispersion in the sense that Montle did remove the influence of observational error in each individual determination of Z. He constructed a calibration of $\langle Z^2 \rangle^{1/2}$ versus age, as he did with $\langle z \rangle$, but from different sources, primarily from the data compiled by Delhaye (1965). The accidental uncertainty in this calibration was only about ± 1 km/sec. Entering the observed $\langle Z^2 \rangle^{1/2}$ into the calibration again yielded an age of 2 X 10^8 years, and considering the uncertainties of both indicated an age in the range 0.5 - 3 x 10^8 years.

Montle's $\langle z \rangle$ and $\langle Z^2 \rangle^{1/2}$ values were determined from different combinations of several different observed quantities (distance, galactic latitude and longitude, systemic radial velocity, and proper motion), though admittedly they both made use of the distance. Moreover, their calibrations versus age were based on different bodies of published data. Therefore, since the two different approaches indicated the same age (2 X 10^8 years) within small ranges, the agreement would seem to indicate that Montle's result is reliable. There is, however, a newer calibration of velocity dispersion versus age by Wielen (1974) which, with Montle's value of 10.0 km/sec, would indicate a considerably older age: around 2 X 10^9 years.

To my knowledge no RS CVn binary is known or suspected of being a member of a star cluster. Fortunately at least one RS CVn binary is known to have a visual companion. Hall (1975c) has shown that the less massive F8 V visual companion to the eclipsing system WW Dra is evidence that WW Dra cannot be pre-main-sequence. The nearest associations are about 1 kpc away whereas the known RS CVn binaries have distances between 50 and 315 pc. Thus, although in principle there could be RS CVn binaries connected with associations, the presently known sample cannot possibly be.

Explanation of Tables 1, 2, 3.

Numbers in parentheses refer to the bibliography, which is numbered correspondingly.

Letters in parentheses refer to notes, which are given at the end of these three tables.

A check mark (√) or a question mark (?) means that the characteristic is present or is possibly present.

h = hotter component

c = cooler component

b = both components

δ_{uv} = ultraviolet excess

δ_{ir} = infrared excess

Corr. = the alternate period variations are correlated with the wave migration

The amplitude of the wave is measured from maximum to minimum, usually in the visual. A range indicates that the amplitude is variable.

P(migr.), used in Tables 1 and 3, is the amount of time required for the wave to complete one retrograde migration and return to the same orbital phase. A range indicates that the migration rate is not uniform.

P(wave), used in Table 2, is the period of the wave-like variation itself.

The distances are taken from Montle (1973).

Table 1. The RS CVn Group

Name	P(orb) days	Spectral Class hot + cool	H & K em.	Hα em.
UX Ari	6.438	G5V + KOIV (28)	c (28)	√ (128)
CQ Aur	10.621	GO (112, C)	c (112)	
SS Boo	7.606	dG5 + dG8 (11)	c (11)	
SS Cam	4.824	dF5 + gG1 (67)	c (102)	
RU Cnc	10.173	dF9 + dG9 (11)	c (11)	
RS CVn	4.798	F4V-IV + KOIV (109)	c (109)	√ (109, 128)
AD Cap	6.118 (H)	G5 (112, C)	b (112)	
UX Com	3.642	G5-9 (53, F)	√ (114)	
RT CrB	5.117	GO (B)	c (102)	
WW Dra	4.630	sgG2 + sgKO (11)	c (11)	√ (110)
Z Her	3.993	F4V-IV + KOIV (106)	c (11)	√ (128)
AW Her	8.801	G2IV + sgK2 (107)	c (11)	
MM Her	7.960	G8IV (70, C)	c (70)	
PW Her	2.881	GO (112, C)	c (112)	
GK Hya	3.587	G4 (B)	c (102)	
RT Lac	5.074	sgG9 + sgK1 (11)	b (11)	√ (110)
AR Lac	1.983	G2IV + KOIV (34)	b (11)	√ (128)
RV Lib	10.722	G5 + K5 (11, 47)	h (102)	
VV Mon	6.051	GO (108, C)	c (102)	
LX Per	8.038	GOV + KOIV (127)	c (127)	√ (128)
SZ Psc	3.966	F8V + K1V-IV (9)	c (9)	√ (128)
TY Pyx	3.199 (2)	G5 + G5 (2)	b (2)	
RW UMa	7.328	dF9 + K1IV (95)	c (11)	
HD 118216	2.613 (38)	F2IV + K IV (38)	c (38)	√ (38)

Table 1 continued. The RS CVn Group

Name	Wave magnitudes	P(migr.) years	Irreg. Lt. C. Var.
UX Ari	0.03 - 0.10 (43)		√ (43)
CQ Aur			
SS Boo	0.05 - 0.19 (62,102)	7.5 (102, J)	
SS Cam	0.11 (102)	78 (5, K)	√ (6)
RU Cnc			
RS CVn	0.05 - 0.20 (57)	8 - 12 (57)	
AD Cap			
UX Com	√ (114)		
RT CrB			
WW Dra	0.06 (102)	√ (102,77)	
Z Her	0.03 (102)	5.1 (102, I)	
AW Her			
MM Her			
PW Her	√ (114)		? (63)
GK Hya	√ (114)		
RT Lac	0.01 - 0.17 (61)	5 - 40 (61)	√ (61)
AR Lac	0.04 (35)	15 - 45 (36)	√ (35)
RV Lib	√ (114)		
VV Mon			
LX Per			
SZ Psc	√ (102)		? (9)
TY Pyx	√ (103)		√ (2)
RW UMa	0.11 (102)	√ (102)	
HD 118216			

Table 1 continued. The RS CVn Group

Name	δ_{uv}	δ_{ir}	Radio em.	**Var.** P(orb.)	Corr.	dist. pc
UX Ari		√ (64)	√ (50)			50
CQ Aur						220
SS Boo	√ (102)					220
SS Cam	? (102)			√ (5)	√ (5)	255
RU Cnc						190
RS CVn	√ (102)	√ (7)		√ (57)	√ (57)	145
AD Cap						250
UX Com						
RT CrB				? (74)		
WW Dra	√ (102)					180
Z Her	√ (102)	√ (7)		? (74)		85
AW Her						315
MM Her						190
PW Her						285
GK Hya						
RT Lac	√ (102)	√ (90)	√ (51)	√ (61)	? (61)	205
AR Lac		√ (7)	√ (49)	√ (55)	√ (36)	50
RV Lib						270
VV Mon				? (74)		260
LX Per						145
SZ Psc		√ (7)		√ (65)		100
TY Pyx						85
RW UMa						150
HD 118216						

Table 1 concluded. The RS CVn Group

Name	Masses in solar units hot + cool	Relative Radii hot + cool	Detached
UX Ari	0.63 + 0.71 (28, L)		
CQ Aur			
SS Boo	0.91 + 0.84 (102)		
SS Cam		0.140 + 0.412 (73)	
RU Cnc		0.064 + 0.173 (73)	
RS CVn	1.35 + 1.40 (102)	0.11 + 0.24 (102)	√
AD Cap	0.5: + 1.1: (102)		
UX Com			
RT CrB			
WW Dra	1.4 + 1.4 (102)	0.14 + 0.24 (102)	√
Z Her	1.22 + 1.10 (102)	0.11 + 0.19 (102)	√
AW Her	1.31 + 1.31 (102)		
MM Her	1.22 + 1.19 (70)	0.12 + 0.065 (70)	√
PW Her	1.35 + 1.58 (102)		
GK Hya			
RT Lac	0.8: + 1.7: (102)		
AR Lac	1.31 + 1.32 (102)	0.19 + 0.32 (102)	√
RV Lib			
VV Mon			
LX Per	1.33 + 1.39 (127, M)		
SZ Psc	1.33 + 1.65 (102)	0.10 + 0.26 (102)	√
TY Pyx	1.20 + 1.22 (2)	0.135 + 0.135 (2)	√
RW UMa	1.1: + 1.1: (102)	0.07 + 0.20 (102)	√
HD 118216			

Table 2. The Long-Period Group

Name	P(orb.) days	Spectral Class hot + cool	H and K emission
ζ And	17.769	K1II (11 , D)	c (11 , E)
λ And	20.521	G8IV-III (11 , D)	c (11 , E)
α Aur	104.023	G0III + G5III (18)	c (11)
12 Cam	80.174	K0III (1 , D)	c (1 , E)
RZ Cnc	21.643	K1III + K4III (26)	h (26)
RZ Eri	39.283	A5-F5V + sgG8 (110)	c (11)
σ Gem	19.605	K1III (11 , D)	c (11 , E)
HK Lac	24.428 (52)	FIV + K0III (18)	c (52)
AR Mon	21.207	F-G + K0II (117)	c (117)
ε UMi	39.481	dA8-dF0 + G5III (11 , 69)	c (11)
BS 7275	28.59	K1IV (A , D)	c (11 , E)
BS 7428	108.58	A + K2III-II (11)	c (11)
BS 8703	24.65	K1IV-IIIp (66, D)	c (66, E)
HD 158393	30.9 (83)	G8III (83, D)	c (83, E)
HD 213389	17.755	K2IV-IIIp (66, D)	c (66, E)

Table 2 continued. The Long-Period Group

Name	Wave magn.	P(wave) days	Distance parsecs
ζ And	0.02 (118)	= P(orb.) (66)	
λ And	0.3 (66)	55.82 (66)	
α Aur	$\surd$ (18)	366 (18)	
12 Cam			
RZ Cnc	0.01 (26)		310
RZ Eri			105
σ Gem	? (66)		
HK Lac	0.10 (18)	25.3 (18)	150
AR Mon			495
ϵ UMi	? (69)		
BS 7275	? (66)		
BS 7428			
BS 8703	0.16 (66)	~100 (66)	
HD 158393			
HD 213389	0.13 (66)	= P(orb.) (66)	

Table 2 concluded. The Long-Period Group

Name	Masses in solar units hot + cool	Radii in solar units hot + cool
ζ And		
λ And		
α Aur	3.03 + 2.91 (135)	
12 Cam		
RZ Cnc	3.1 + 0.55 (26)	11 + 13 (26)
RZ Eri	2.23 + 1.72 (102)	
σ Gem		
HK Lac		
AR Mon	2.6 + 0.8 (113)	6 + 15 (113)
ε UMi		
BS 7275		
BS 7428		
BS 8703		
HD 158393		
HD 213389		

Table 3. The Short-Period Group

Name	P(orb.) days	Spectral Class hot + cool	H and K emission
RT And	0.629	F8V + G5V (40)	√ (79, G)
SV Cam	0.593	G3V-IV + K3V (46)	√ (68, G)
WY Cnc	0.829	G± (B, F)	h (102)
CG Cyg	0.631	G9V-IV (89, F)	√ (89, G)
UV Psc	0.861	G2 (112, C)	b (112)
ER Vul	0.698	G0V + G5V (98)	√ (20, G)

Table 3 continued. The Short-Period Group

Name	Wave magn.	P(migr.) years	Irregular Lt.C.Var.	Distance parsecs
RT And			√ (40)	95
SV Cam			√ (125)	
WY Cnc	0.02 (33)	5.5: (102)	√ (102)	160
CG Cyg	0.07 (88)	10 (59)		
UV Psc	0.04 (102)			125
ER Vul			√ (98)	45

Table 3 continued. The Short-Period Group

Name	δ_{uv}	δ_{ir}	Var. P(orb.)	Corr.
RT And			√ (131)	
SV Cam			√ (46)	
WY Cnc				
CG Cyg		√ (90)	√ (59)	√ (59)
UV Psc	? (102)			
ER Vul				

Table 3 concluded. The Short-Period Group

Name	Masses in solar units hot + cool	Relative Radii hot + cool
RT And	1.50 + 0.99 (40)	0.322 + 0.239 (40)
SV Cam		0.40 + 0.25 (46)
WY Cnc		0.090 + 0.303 (73)
CG Cyg		
UV Psc		
ER Vul	1.07 + 0.98 (98)	0.297 + 0.282 (73)

Notes to Tables 1, 2, 3.

a. the 1964 Yale Bright Star Catalogue

b. the 1969 General Catalogue of Variable Stars

c. spectral type of the hotter component

d. only this component seen in the spectrum

e. assuming the one component seen in the spectrum is the cooler

f. refers to the composite spectrum

g. not known which component(s) responsible for the emission

h. Oliver (1974a) says the period is around 3 days

i. 3.8 years is also possible

j. 9.3 years and 6.4 years are also possible

k. based on 39 years for half a migration cycle

l. $M \sin^3 i$ values

m. $M \sin^3 i$ values, but $i \approx 90^\circ$

7. Summary of Properties

In this final subsection I want to recapitulate by listing in Table 4 the observed properties and physical characteristics of the RS CVn binaries and indicating what fraction of the known sample exhibits (or is known to exhibit) each. The first three are exhibited by all 24 because they make up my proposed working definition. The denominator in some of the fractions is less than 24 because sufficient data are not available for all. In evaluating the numerator I have counted question-mark entries in Table 1 as one half. Percentages are given as lower limits if the accumulation of additional data is likely to increase the value.

Should we add additional membership requirements to make the sample of RS CVn binaries more pure and homogeneous? I think this is premature. In most cases the absence of a given property in a certain binary is not firmly established; there is inadequate observation to decide. In cases where apparently adequate observations are available but a certain property such as a wave-like distortion or infrared excess is not present, it might be that the binary is near the minimum of its "sunspot cycle" and only temporarily not displaying those properties. Based on more permanent features like spectral type and mass ratio one might be tempted to purge certain members. Mentioned earlier was TY Pyx, the only binary not having a cooler star near K0 IV; and the unit mass ratio criterion would eliminate RT Lac and RV Lib. Probably one should consider the conservative use of sub-groups rather than exclude certain binaries outright. For the near future, however, I think the first three properties are sufficient to define an "RS CVn binary", to allow us to make use of existing data in theoretical investigations, and to guide the search for additional members.

Table 4. Summary of Properties for the RS CVn Group

	Property	Fraction	
1.	Orbital period between 1 day and 2 weeks	24/24	100 %
2.	Strong H & K emission seen outside eclipse	24/24	100 %
3.	Hotter star is F or G, V or IV	24/24	100 %
4.	H & K emission is from cooler star (or both)	22/23	96 %
5.	Cooler star is around KO IV	23/24	96 %
6.	Hα emission seen outside eclipse	6/6	100 %
7.	Wave-like distortion outside eclipse	> 15/24	> 62 %
8.	Wave migrates towards decreasing phase	> 7/15	> 47 %
9.	Variable depth of primary minimum		
10.	Variable displacement of secondary eclipse		
11.	Irregular light curve variations	> 6/24	> 25 %
12.	UV excess in one or both components	5.5/12	46 %
13.	IR excess in one or both components	5/6	83 %
14.	Radio emission	> 3/24	> 12 %
15.	Variable orbital period	> 6.5/24	> 27 %
16.	Period variations correlated with migration	> 3.5/6.5	> 54 %
17.	Mass ratio near unity	13/15	87 %
18.	Binary is detached	8/8	100 %

III. THE RELATED BINARIES

A. The Long-Period Group

As explained earlier, I am defining this group as binaries with periods greater than 2 weeks in which one component is of spectral class G-K IV-II and displays strong H and K emission. Table 2 lists all 15 candidates for this group which I am aware of, but there may be more.

Only 3 of these 15 have two-lined spectra, and only 4 are known to be eclipsing. In the 4 cases where something is known about the hotter component, we find its spectral type ranging between A-F and early K.

The mass ratio has been measured only in 4 systems, but in 3 it is far from unity, with the cooler component always the less massive. This would seem to set them apart from the RS CVn group, where the mass ratio is almost always very near unity and, in the two exceptions, the cooler component is the more massive.

An important question to answer is whether binaries of this group are detached or semi-detached. Broglia and Conconi (1973) found RZ Cnc semi-detached. The data in Table 2 indicate that AR Mon is also semi-detached. According to Lloyd Evans (1973) the primary component in HD 158393 and other one-spectrum members of this group is "probably close to filling its Roche lobe". This indication that binaries of the long-period group are semi-detached is another property which seems to set them apart from the RS CVn binaries.

Binaries of the long-period group do display wave-like distortions in their light curves. The amplitudes of these waves are comparable to those displayed by the RS CVn binaries. But we do not as a rule see them migrating slowly towards decreasing orbital phase as we do in the RS CVn binaries. According to the model of Hall (1972), the key to the migration phenomenon is the binary's attempt to reconcile synchronous rotation and differential rotation. Thus it seems the cooler star in binaries of the long-period group is not as a rule rotating synchronously, whereas the cooler star in the RS CVn binaries always is. According to Herbst (1973) the period of the wave in ζ And and HD 213389 equals the orbital period, whereas in λ And, α Aur, and BS 8703 the period of the wave is longer by a factor of several. The wave period in HK Lac is also longer than the orbital period, but only by about 5%. This would correspond to a migration period of only about 2 years, but in the direction of *increasing* orbital phase. Lloyd Evans (1973) finds the cooler star in HD 158393 rotating rapidly. Thus we see 2 rotating synchronously, 3 rotating much more slowly, 1 rotating a little more slowly, and 1 rotating more rapidly. The reason for this general lack of synchronism is not entirely obvious. Although their orbital periods are longer, the star in question is also larger and the tidal forces felt by it might be comparable to those felt by the KO IV star in the RS CVn group.

Masses have been determined only for α Aur, RZ Eri, and AR Mon. These indicate total masses in the range 3.2 $M_{\odot}$ to 5.94 $M_{\odot}$, which is entirely above the upper end of the RS CVn mass range. This fact is, indirectly, consistent with the Period-Luminosity relation noted by Gratton (1950).

B. The Short-Period Group

As mentioned earlier, I am defining this group as non-contact binaries with periods less than 1 day in which the hotter component is of spectral class F-G V-IV and H and K emission is displayed in one or both components. Table 3 lists 6 candidates for this group which I am aware of. XY UMa, discussed recently by Geyer (1976), appears to be another. The requirement that they not be contact is necessary to separate this group from the W UMa binaries.

Oliver (1974a) says that the H and K emission arises from the hotter star in WY Cnc and from both in UV Psc; in the other 4 it is not known which star is responsible. Thus no clear pattern can be seen here.

There is a wave in WY Cnc, CG Cyg, UV Psc, and XY UMa; the waves in the first two have been found to migrate towards decreasing orbital phase. Irregular light curve variations, of the sort described with the RS CVn binaries, are observed dramatically in RT And, SV Cam, WY Cnc, and ER Vul.

Large alternate period variations, $\Delta P/P \sim 10^{-5}$, are observed in RT And, SV Cam, and CG Cyg. In CG Cyg, the one of these with a well-defined wave, the period variations seem to be correlated with the wave migration (Hall 1975b).

This group seems to be closely related to the RS CVn group if we judge by similarity of observed properties. For those binaries like RT And, SV Cam, and ER Vul which show only irregular light curve changes instead of a persistent wave, it might be that large spotted regions are appearing and disappearing without, as far as we can tell, remembering to have a preferential longitude.

The general shape of the light curve outside eclipse indicates that binaries in this group are neither contact nor supercontact. With reliable absolute dimensions available for only a few, it is difficult to be sure whether these binaries are detached or semi-detached. It seems clear that ER Vul is detached (Northcott and Bakos 1967) and it is possible that RT And is detached also (Dean 1974) even though Kopal (1959, Table 7-5) considered it semi-detached. Frieboes-Conde and Herczeg (1973) think SV Cam is probably semi-detached, with the hotter component filling its Roche lobe.

The range of total mass, judging by just RT And and ER Vul, is included within the RS CVn mass range. The mass ratio for ER Vul is near unity (Northcott and Bakos 1967); the mass ratio for RT And is not reliably determined but might be around 1.5 (Dean 1974). I think it is premature to decide now whether or not the short-period group and the RS CVn group are similar with respect to mass ratio.

This short-period group seems to be a subset of the short-period eclipsing binaries with β Lyrae-type light curves,discussed recently by Lucy (1975). Lucy explains that, although detached binaries shortly before they undergo mass exchange have primaries nearly filling their Roche lobes and hence will have β Lyrae-type light curves (e.g. MR Cyg) and although some semi-detached binaries actually undergoing post-main-sequence mass exchange have β Lyrae-type light curves (e.g. β Lyr itself), most eclipsing binaries with β Lyrae-type light curves have periods less than 1 day. Lucy further notes a sharp drop in frequency of these binaries with orbital periods below $0^{d}.45$ which coincides with a sharp rise in the frequency of the W UMa systems.

This group is defined as binaries in which the hotter component is a dKe or a dMe star. Here the emission refers to strong H and K emission. Three well-known examples of such a group would be BY Dra, CC Eri, and YY Gem, but there are certainly others which would fit such a definition. See, for example, Krzeminski (1969, Table I).

In all three of these binaries, both components are known to exhibit the H and K emission. YY Gem is similar to the RS CVn binaries in that the radial velocity curve amplitude and shape are virtually the same for both the absorption lines and the emission lines (although there is a systematic difference in γ velocity) and the doppler width of the emission lines corresponds to synchronous rotation (Bopp 1974a). All three are also known to exhibit Hα emission (Anderson and Bopp 1975, Evans 1971, Moffett and Bopp 1971).

All three are known to exhibit wave-like distortions in their light curves. In BY Dra and CC Eri the wave is remarkably well-defined (Bopp and Evans 1973) and in YY Gem a wave seems definitely to be present (Budding 1976). In the first two the wave is known to have changed phase with respect to the orbital period, and in the one case of BY Dra the period of this wave has been determined and found to be $3\overset{d}{.}838$. Since the orbital period of BY Dra is $5\overset{d}{.}976$, its wave would migrate towards decreasing orbital phase, as is observed in the RS CVn binaries, but very rapidly. It is clear that, unlike the RS CVn binaries, BY Dra is far from synchronous rotation. The amplitude of the wave in CC Eri and BY Dra is known to be variable, another similarity between this group and the RS CVn group. In fact one could say that the amplitude can dwindle to zero, because at times observations of both CC Eri (Evans 1971) and BY Dra (Martins 1975) have failed to reveal any measurable light variation. Bopp (1974a) summarizes by saying that spots and active regions on these stars apparently can develop and disappear on time scales of one month (YY Gem) or a few months (CC Eri and BY Dra).

All three undergo rapid flare activity (Cristaldi and Rodono 1971, Nather and Harwood 1972, Moffett and Bopp 1971). In other words, all three are flare stars. It is quite possible that flares of the same intrinsic brightness occur in the RS CVn binaries also, but that these flares go unnoticed because the RS CVn binaries are intrinsically more luminous than the dMe or dKe binaries.

It has been suspected that the orbital period of YY Gem is variable. The other two are not eclipsing variables and a variation in the orbital period will be much more difficult to detect and substantiate. The detection of circumstellar material in YY Gem (Bopp 1974 a) suggests that matter is actually being ejected from one or both components. Such mass ejection would make YY Gem similar to the RS CVn binaries if mass ejection is indeed the cause of their period changes.

Although radio emission has not yet been detected from any of these three flare star binaries, it is well known that many flare stars do emit radio waves coincident in time with optical flares. Thus radio emission might be another property which the flare star group and the RS CVn group have in common. Bopp, Gehrz, and Hackwell (1974), however, observed BY Dra and YY Gem and found no measurable infrared excess.

The mass ratio in YY Gem is exactly unity within the observational uncertainty. That of BY Dra is rather close to unity. Very few masses are known for binaries of this group, but the total mass of each binary system is no doubt considerably less than that for each RS CVn binary. Their orbital periods range from $0\overset{d}{.}8$ to $6\overset{d}{.}0$, overlapping somewhat the range for the RS CVn group but tending to be somewhat shorter.

It is generally agreed that there are important differences between the dwarf M and dwarf K stars which have H and K emission and those which do not. As Krzeminski (1969) points out, periodic or quasi-periodic light variations are found in some dMe stars, but never in dM stars.

D. V471 Tau = BD +16°516

V471 Tau is a detached binary consisting of a KO dwarf and a white dwarf and having an orbital period of 12.5 hours. No similar binary is yet known. The work of Flesch, Oliver, and Smak (1974) first made me aware of the many properties it shares with the RS CVn binaries.

The cooler component displays strong H and K emission. There is a well-defined wave, attributable to the cooler star, which migrates towards decreasing orbital phase with a migration period of around one year. The orbital period is variable (Lohsen 1975) in a way which might possibly be correlated with the wave migration in the sense found in the RS CVn binaries (Oliver 1975). The difficulty in establishing such a correlation is that the migration period is nearly a year and the orbital period is nearly a half day.

The mass ratio is very nearly unity according to Young and Nelson (1972), yet another point of similarity with the RS CVn binaries.

E. The W UMa Group

In this group I am considering W UMa binaries in which there is photometric evidence of uneven surface brightness distribution on one or both components. Among these are U Peg (Binnendijk 1960a), AH Vir, (Binnendijk 1960b), RZ Com (Binnendijk 1964), SW Lac (Bookmeyer 1965), AM Leo (Binnendijk 1969), VW Cep (Leung and Jurkevich 1969), W UMa (Rigterink 1972), and 44 ι Boo (Bergeat et al. 1972). There may be others which would fit into this group.

Binnendijk analyzed light curves of AH Vir, U Peg, RZ Com, and AM Leo, paying particular attention to asymmetries and changes in the light curve as a function of time. Bookmeyer did the same for SW Lac, and Rigterink did for W UMa. In every case they arrived at a model in which there was a subluminous region on the surface of the larger = more massive star. This subluminous region caused the affected hemisphere to be fainter than the other hemisphere by as much as $0\overset{m}{.}05$. In the case of AH Vir and SW Lac there was a significant displacement of the phase of secondary minimum with respect to primary minimum. In both cases this could be explained as a direct consequence of the subluminous region; this is, of course, the same way Hall (1972) accounted for the displaced secondary minimum in RS CVn.

In the case of U Peg, SW Lac, and W UMa the orientation of the subluminous region was different in different epochs. Since the subluminous region would produce a wave-like distortion in the light curve, this would suggest a migration effect such as observed in the RS CVn binaries. In fact Rigterink shows that the wave in W UMa is migrating with a period of about 500 days; and Leung and Jurkevich show that there is a wave in VW Cep migrating through its light curve with a period of 718 days. In both cases the migration is towards decreasing orbital phase. Thus the similarity between the RS CVn group and this W UMa group is remarkable.

In addition to the effects attributed to a subluminous region on the surface of one star, large changes in the overall heights of the light curve maxima are observed in SW Lac by Bookmeyer.

Most (6 out of 8) of these W UMa systems are known to undergo large period changes. Despite the fact that half of these binaries are known to have visual companions (AH Vir, AM Leo, W UMa, and 44 ι Boo), it is highly unlikely that orbital motion around a third body can entirely account for the observed period variations. This is because the period variations are very large ($\Delta P/P \sim 10^{-5}$), alternate in sign with a time scale of $\sim$10 years, and are often quite abrupt. One expects that a migrating wave will distort the shape of both eclipses and thereby generate spurious O-C changes which do not represent true period changes. Van't Veer (1973) has demonstrated very nicely that this is happening in VW Cep. But it is clear that these spurious changes are only a small part of the overall O-C changes; hence real period changes are occurring, not only in VW Cep but also in the others. In the case of AH Vir, Binnendijk discussed possible explanations for the observed period variations in terms of mass ejection from the subluminous region, but he was not able to reach a definite conclusion. In the case of W UMa Binnendijk (1966) showed that the flare observed by Kuhi (1964) coincided in time with a large abrupt period increase.

In AH Vir, RZ Com, SW Lac, and AM Leo there were epochs when the subluminous region was absent; at least there were epochs when the light curve appeared normal. This come-and-go nature of the subluminous region suggests something like the "sunspot cycle" which Hall (1972) proposed to explain the variable wave amplitude in RS CVn and which might explain the variable wave amplitudes in the flare star binaries also. In this connection the discussion of 44 ι Boo by Bergeat et al. (1972) is very interesting. They concluded that sudden changes in the orbital period and various irregularities in the light curve indicated the existence of active and quiet epochs, with the interval between active epochs being around 10 years. Light curve characteristics considered by them were (1) unequal maxima, (2) displaced secondary minima, (3) general irregular appearance, and (4) a temporal enhancement or reduction of the light curve occurring at certain phases. They pointed out the similarity between this interval and the 11-year solar sunspot cycle.

H and K emission has been observed in at least one of these systems, namely W UMa itself (Struve and Horak 1950). I am not sure whether or not this will turn out to be a common property of the group. Koch (1974) concludes that his narrow band CN observations indicate an overabundance in the W UMa binaries (which he refers to as strongly interacting binaries), but I am not sure what the ultimate implication of this observation will prove to be.

Milone (1976) has detected very strong infrared excess in one W UMa-type binary, RW Com. This suggests that perhaps infrared excess might prove to be another property which they have in common with the RS CVn binaries.

IV. CONCLUDING REMARKS

A. How to Understand these Properties

Table 5 is provided to illustrate the degree to which binaries of the RS CVn group and the five related groups display similar properties. Such a summary must be somewhat subjective since all of the members of a given group do not always display a given property. A "yes" indicates that all or most do or that accumulation of additional data will probably indicate that they do. A "maybe" indicates that the property is possibly characteristic of that group. A question mark means that the situation is ambiguous at present. Particulars about any entry can be found in the text or in Tables 1, 2, and 3. It is remarkable that several of the properties occur in all or most of the groups. I feel that, with the accumulation of more data, these similarities will become even more apparent.

I think it is almost certain now that the many peculiar properties reviewed in this paper are to be understood with a picture of strong chromospheric activity and large spotted regions on convective stars in binary systems. As indicated throughout this review, such a picture can account for the strong H and K emission, the Hα emission, the correlation of both these types of emission with phase, the persistent wave outside eclipse, the migration of this wave towards decreasing orbital phase (when the spotted star is rotating synchronously), the variable displacement of secondary eclipse and the variable depth of primary minimum and the correlation of both these variations with wave migration, the anomalously shallow depth of secondary minimum, the nonconstancy of the migration rate and the variable ampli-

Table 5. Summary of Properties for the Six Groups

Property	RS CVn	L-P	S-P	W UMa	Flare	V471
a. H and K emission	yes	yes	yes	maybe	yes	yes
b. H α emission	yes				yes	
c. Wave	yes	yes	yes	yes	yes	yes
d. Displaced secondary	yes			yes		
e. Synchronism	yes	no	yes	yes	no	yes
f. Retrograde migration	yes	—	yes	yes	—	yes
g. Evidence of a spot cycle	yes			yes	yes	
h. Irregular light curve variations	yes		yes	yes		yes
i. Flare activity				maybe	yes	
j. UV excess	yes		maybe		yes	
k. IR excess	yes		yes	maybe		
l. Radio emission	yes				yes	
m. Variable orbital period	yes		yes	yes	maybe	yes
n. Period variations correlated with migration	yes		yes	maybe		maybe
o. Mass ratio near unity	yes	no	maybe	no	yes	yes
p. d, sd, or c	d	sd	?	c	d	d

tude of the wave and the correlation of these two with each other, the period changes and their correlation with wave migration, the depressions in some light curves around first and fourth contact of primary eclipse, the irregular light curve changes, flare activity, ultraviolet excess, infrared excess, and radio emission.

The spot model proposed by Hall (1972) for RS CVn is remarkably similar to the spot model developed by Torres and Ferraz Mello (1973) for AU Mic, by Bopp and Evans (1973) for BY Dra and CC Eri, and elaborated upon by Vogt (1975). Mullen (1974) has shown that starspots can be derived from his convection cell hypothesis which are in fair agreement with the size and temperature of those observed by Bopp and Evans (1973). He further shows that efficient dynamo action is a possible mechanism for generating the required large surface fields and that tidal effects may influence starspot formation. Another interesting paper is that of Worden (1974). Although both Mullen and Worden were interested mainly in the flare stars, their results apply to any late-type star which has a deep convective envelope and hence should be very useful in understanding spot activity in the RS CVn group, the long-period group, the short-period group, the W UMa group, and V471 Tau.

In my opinion the case for strong chromospheric activity and for large spotted regions is very convincing. It should be pointed out, however, that other explanations have been proposed by Evans (1971), by Catalano and Rodono (1967, 1969, and 1974), and by Ulrich (1976). Evans since then has said (Bopp and Evans 1973) that he now believes the spot model is on the right track. Many people object to a spot model on grounds that it is methodologically distasteful, but this objection does not disprove the existence of spots.

The extreme degree of the chromospheric activity and spottedness might mean that the affected stars are thermally unstable. But, even though the binaries in these various groups may have thermal instability in common, it would not necessarily follow that they got this thermal instability as a result of similar evolutionary histories.

B. The Evolutionary Status of the Related Groups

Figure 1 shows that the various groups, as defined in this paper, occupy different regions in the total-mass versus orbital-period plane, but I am not sure what the significance of this segregation is. It is conceivable that binaries of the long-period group are the long-period counterparts of the RS CVn binaries, that binaries of the short-period group are the short-period counterparts, or that the flare stars are the low-mass equivalents. But it is quite possible that no such connection is true.

Biermann and Hall (1976) have discussed the problem of the evolutionary status of the RS CVn binaries. After considering that they might be pre-main-sequence or post-main-sequence and that their main-sequence counterparts might be either binaries or single stars, they felt the most likely explanation was that the RS CVn binaries are in a thermal phase following fission of a rapidly rotating main-sequence single star. Ulrich (1976) argues, however, that the RS CVn binaries in fact can be understood as products of post-main-sequence evolution of a binary system. This matter has not been settled yet.

Koch (1970) suggested that RZ Cnc (which I include in the long-period group) is in pre-main-sequence contraction. If so, it would be thermally unstable but not for the same reason Biermann and Hall

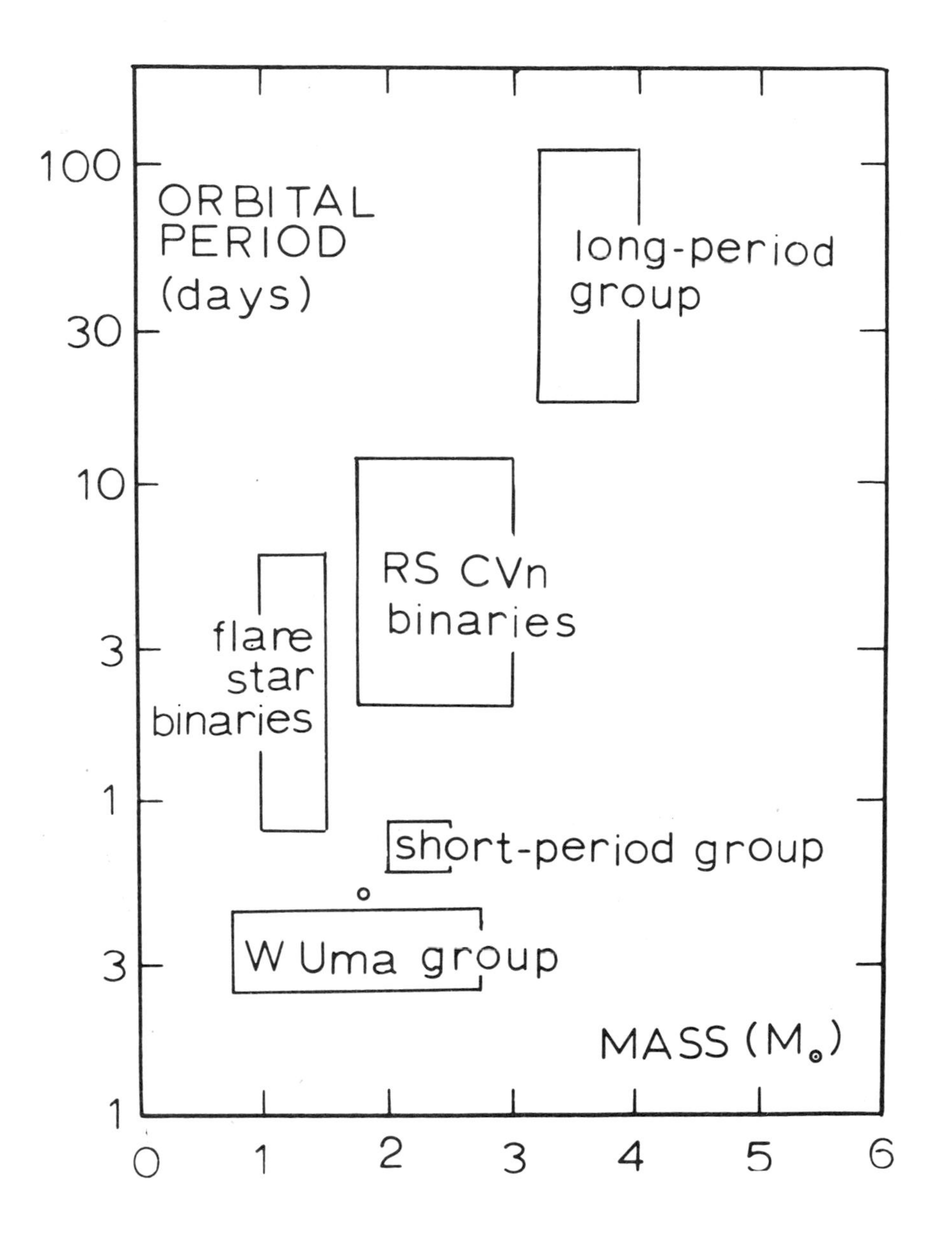

Figure 1. The regions occupied by the six binary star groups in the orbital-period versus total-mass plane. The small circle represents V471 Tau.

suggested the RS CVn binaries might be. Other people have considered binaries in the long-period group to be the result of post-main-sequence evolution of a close binary. This matter has not been settled yet. Recent observations of Bolton (1975) argue against a pre-main-sequence interpretation. He obtained 16 Å/mm red spectrograms of several stars of this group and compared them visually with spectrograms of stars with known Li line strengths. The indication was that binaries of the long-period group do not have the very strong Li lines characteristic of young stars of the same spectral type.

Wilson and Woolley (1970) examined a large number of dwarf K and dwarf M stars and found a clear correlation between the strength of H and K reversal and various parameters of their galactic orbits. From this they concluded that the dKe and dMe stars are young, whereas the dK and dM stars are old. Bopp (1974b) has detected Li in one dMe star.

It is significant that 7 of the 8 binaries in my W UMa group are known to be of type W, and the one exception (AM Leo) is not yet classified (Rucinski 1974). According to Rucinski the W-type, in contrast to the A-type, are thermally unstable. According to Lucy (1975) the W-type W UMa binaries _are_ evolving on a nuclear time scale but, due to an inability to attain structures in thermal equilibrium, they are condemned to undergo thermal relaxation oscillations about a state of marginal contact. If so, then we have yet another reason for the thermal instability.

For the evolution of V471 Tau, Nelson (1976) suggested fission of a rapidly rotating main-sequence star, while others at the same symposium suggested the spiral-in of a post-main-sequence binary which originally had a much longer period. There is no consensus on this matter yet.

An answer to the question of a possible evolutionary connection between the groups should involve careful consideration of space densities, mass ranges and ratios, angular momentum, luminosity fuctions, relative ages, kinematical properties, and any real abundance anomalies.

C. Comparing Migration Rates

It is interesting to examine the ratio P(migr.)/P(orb.). Excluding those examples where it appears we do not have synchronous rotation, we are left with SS Boo, SS Cam, RS CVn, Z Her, RT Lac, AR Lac, WY Cnc, CG Cyg, V471 Tau, VW Cep, and W UMa. The range for the RS CVn group is about 350 to 6000. The two examples in the short-period group range between about 2400 and 5800. V471 Tau is around 700. The two examples in the W UMa group range between 1500 and 2600. Thus, with no exceptions, binaries in the related groups fall within the range of P(migr.)/P(orb.) shown by the RS CVn group.

It is particularly interesting that binaries of the W UMa group fit also. They are contact (or supercontact) binaries whereas most of those in the other groups are certain to be detached.

D. Explaining the Amplitude of the O-C Variations

There is a simple way to understand why some of the binaries in these groups have such extraordinarily large O-C variations and, at the same time, to check the basic idea that mass loss is causing the period changes.

Let us assume, for the purposes of this discussion, that the relative mass outflow per year is roughly the same for binaries in all of the different groups. Further let us assume that, whatever the detailed mechanism for the resultant period changes proves to be, the coefficient relating $d \ln P / dt$ with $d \ln M / dt$ is also roughly the same. With these two assumptions $d \ln P / dt$, which is the degree of curvature in the O-C curve, is roughly the same for all. It can then be shown that the semi-amplitude of the up-and-down deviations in the O-C curve is determined only by the migration period, the relation being

$$\Delta(O-C) : P^2(\text{migr.}) . \qquad (1)$$

This relation can be checked by using data and sources already referred to in this review and plotting log Δ(O-C) versus log P(migr.) in Figure 2. No period variation has been detected in 3 of the systems, but we can place an upper limit on the amplitude of the O-C variation. Because the migration rates are not constant in all, their periods have been indicated with a range. The straight line has a slope of 2, as would be required by equation (1). The data adhere to this line surprisingly well, indicating that this simple-minded understanding is on the right track. Let me emphasize what a huge range of O-C variation is hereby accounted for: from Δ(O-C) = ± 17 sec in V471 Tau to Δ(O-C) = ± 0.25 days in SS Cam.

If mass loss is indeed producing the period changes, then there would be some net loss of orbital angular momentum and hence there should be a tendency for the orbital period to decrease. If we look at the overall trend of the O-C curves we see a secular period decrease in RT And, RS CVn, CG Cyg, AR Lac, and possibly Z Her. There is a secular increase in RT Lac. There is no clear trend in SS Cam,

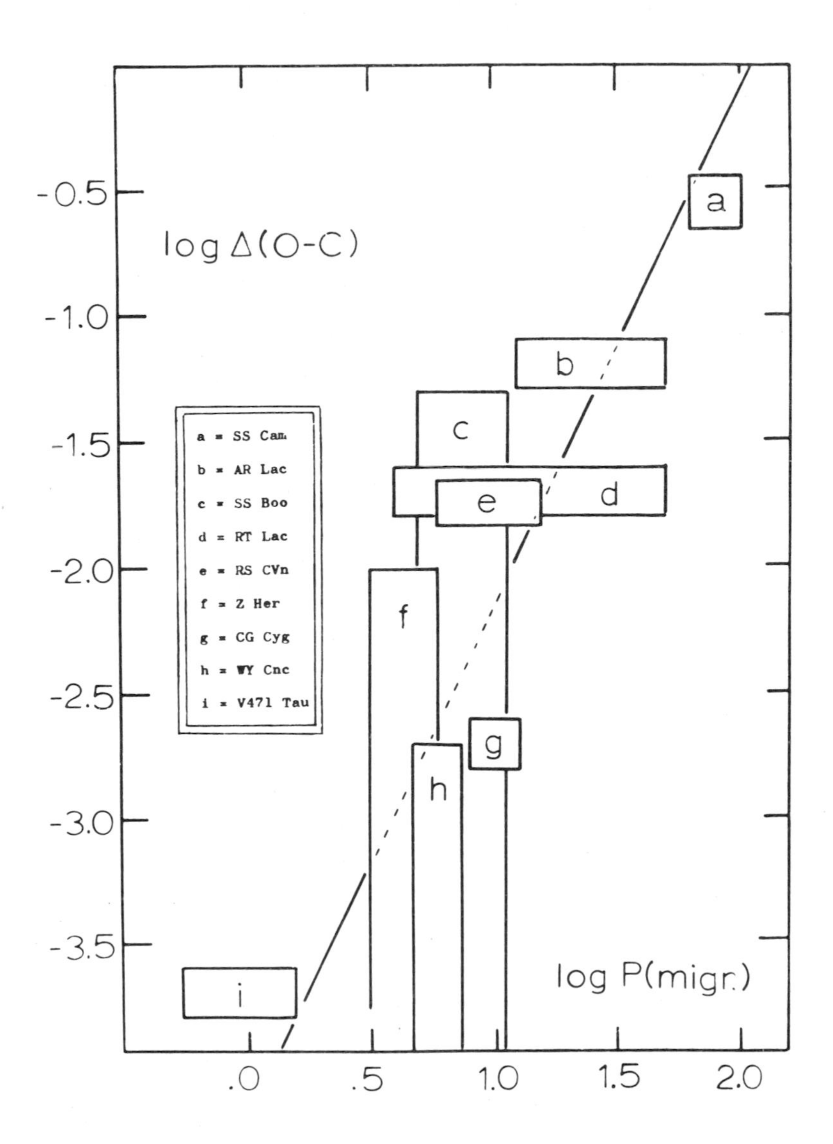

Figure 2. Δ(O-C) is the semi-amplitude of the up-and-down variations in the O-C curve, in days. P(migr.) is the migration period of the distortion wave, in years. Only upper limits to Δ(O-C) are available for SS Boo, Z Her, and WY Cnc. The size of each rectangle includes a nominal ± 10 % accidental uncertainty. The straight line has a slope of 2 and thus corresponds to equation (1).

SV Cam, and V471 Tau. Thus, on the average, there is some indication of the expected net orbital angular momentum loss.

More data would help us understand these interesting period variations better, but it is very difficult to determine the migration rate of an RS CVn binary because one needs light curves fairly complete outside eclipse and distributed without significant gaps over the migration period, which can be decades. Light curves belonging to different migration cycles cannot always be combined in a simple way because the migration rate is not always constant. The amplitude of the wave is usually variable and is sometimes small. And irregular intrinsic light curve changes often complicate the interpretation further. It is even more difficult to correlate wave migration with period changes because, in addition to the migration rate, one must have a well-defined O-C curve covering the same interval of time.

Another difficulty which observers of RS CVn binaries are sadly familiar with is the unfortunate coincidence that many of the orbital periods are near an integral number of days. Of the 24 systems in Table 1, these are AD Cap ($3^{d}.059$), Z Her ($3^{d}.993$), MM Her ($7^{d}.960$), RT Lac ($5^{d}.074$), AR Lac ($1^{d}.983$), VV Mon ($6^{d}.051$), LX Per ($8^{d}.038$), and SZ Psc ($3^{d}.966$). In a random set of irrational numbers greater than unity, fewer than about 12% should have their digits to the right of the decimal greater than 0.960 but less than 0.070. In our sample of 24 we would most reasonably expect about 3. Nature was unkind and gave us 8.

V. Acknowledgements

I acknowledge a grant from the Vanderbilt University Research Council which partially supported my travel to this colloquium. And I am very grateful to Drs. John P. Oliver, Peter Biermann, Edward J. Weiler, and Slavek Rucinski for extremely valuable discussions which helped me considerably in preparing this review.

VI. References

1. Abt, H. A., Dukes, R. J., Weaver, W. B. 1969, Ap. J. 157, 717.

2. Andersen, J., Popper, D. M. 1975, Astr. Astrophys. 39, 131.

3. Anderson, C. M., Bopp, B. W. 1975, B.A.A.S. 7, 235.

4. Arnold, C. N., Hall, D. S. 1973, I.A.U. Inf. Bull. Var. Stars No. 843.

5. Arnold, C. N., Hall, D. S., Montle, R. E. 1973, I.A.U. Inf. Bull. Var. Stars No. 796.

6. Arnold, C. N., Hall, D. S., Montle, R. E. 1975, in preparation.

7. Atkins, H. L., Hall, D. S. 1972, P.A.S.P. 84, 638.

8. Babaev, B. 1971, Astr. Circ. (U.S.S.R.) No. 628, 5.

9. Bakos, G. A., Heard, J. F. 1958, A.J. 63, 302.

10. Bergeat, J., Lunel, M., Sibille, F., van't Veer, F. 1972, Astr. Astrophys. 17, 215.

11. Bidelman, W. P. 1954, Ap. J. Suppl. 1, 175.

12. Biermann, P., Hall, D. S. 1976, I.A.U. Symposium No. 73, in press.

13. Binnendijk, L. 1960a, A.J. 65, 88.

14. Binnendijk, L. 1960b, A.J. 65, 358.

15. Binnendijk, L. 1964, A.J. 69, 154.

16. Binnendijk, L. 1966, A.J. 71, 340.

17. Binnendijk, L. 1969, A.J. 74, 1031.

18. Blanco, C., Catalano, S. 1970, Astr. Astrophys. 4, 482.

19. Bolton, C. T. 1975, private communication.

20. Bond, H. E. 1970, P.A.S.P. 82, 321.

21. Bookmeyer, B. B. 1965, A.J. 70, 415.

22. Bopp, B. W. 1974a, Ap.J. 193, 389.

23. Bopp, B. W. 1974b, P.A.S.P. 86, 281.

24. Bopp, B. W., Evans, D. S. 1973, M.N.R.A.S. 164, 343.

25. Bopp, B. W., Gehrz, R. D., Hackwell, J. A. 1974, P.A.S.P. 86, 989.

26. Broglia, P., Conconi, P. 1973, Mem. Soc. Astr. Ital. 44, 87.

27. Budding, E. 1976, I.A.U. Colloquium No. 29, in press.

28. Carlos, R. C., Popper, D. M. 1971, P.A.S.P. 83, 504.

29. Catalano, S. 1973, I.A.U. Symposium No. 51, p. 61.

30. Catalano, S., Rodono, M. 1967, Mem. Soc. Astr. Ital. 38, 395.

31. Catalano, S., Rodono, M. 1969, in Non-Periodic Phenomena in Variable Stars, ed. L. Detre (Budapest: Academic Press) p. 435.

32. Catalano, S., Rodono, M. 1974, P.A.S.P. 86, 390.

33. Chambliss, C. R. 1965, A.J. 70, 741.

34. Chambliss, C. R. 1974, B.A.A.S. 6, 467.

35. Chambliss, C. R. 1975, private communication.

36. Chambliss, C. R., Hall, D. S., Richardson, T. R. 1975, in preparation.

37. Chisari, D., Lacona, G. 1965, Mem. Astr. Soc. Ital. 36, 463.

38. Conti, P. S. 1967, Ap.J. 149, 629.

39. Cristaldi, S., Rodono, M. 1971, Astr. Astrophys. 12, 152.

40. Dean, C. A. 1974, P.A.S.P. 86, 912.

41. Delhaye, J. 1965, in Galactic Structure, eds. A. Blaauw and M. Schmidt (Chicago: Univ. of Chicago Press) p. 61.

42. Dworak, T. Z. 1973, I.A.U. Inf. Bull. Var. Stars No. 846.

43. Evans, C. R., Hall, D. S. 1974, I.A.U. Inf. Bull. Var. Stars No. 945.

44. Evans, D. S. 1971, M.N.R.A.S. 154, 329.

45. Flesch, T. R., Oliver, J. P., Smak, J. I. 1974, B.A.A.S. 6, 467.

46. Frieboes-Conde, H., Herczeg, T. 1973, Astr. Astrophys. Suppl. 12, 1.

47. Gaposchkin, S. 1953, Ann. Harvard Coll. Obs. **113**, 69.

48. Geyer, E. H. 1976, I.A.U. Colloquium No. 29, in press.

49. Gibson, D. M., Hjellming, R. M. 1974, P.A.S.P. **86**, 652.

50. Gibson, D. M., Hjellming, R. M., Owen, F. N. 1975, Ap.J. **200**, L99.

51. Gibson, R. M., Hjellming, R. M., Owen, F. N. 1975, I.A.U. Circ. No. 2789.

52. Gorza, W. L., Heard, J. F. 1971, Pub. David Dunlap Obs. **3**, 107.

53. Götz, W., Wenzel, W. 1962, Mitt. Veränd. Sterne No. 701.

54. Gratton, L. 1950, Ap.J. **111**, 31.

55. Guarnieri, A., Bonifazi, A., Battistini, P. 1975, Astr. Astrophys. Suppl. **20**, 199.

56. Hall, D. S. 1967, A.J. **72**, 301.

57. Hall, D. S. 1972, P.A.S.P. **84**, 323.

58. Hall, D. S. 1975a, Acta Astr. **25**, 1.

59. Hall, D. S. 1975b, Acta Astr. **25**, No. 3, in press.

60. Hall, D. S. 1975c, Acta Astr. **25**, No. 3, in press.

61. Hall, D. S., Haslag, K. P. 1976, I.A.U. Colloquium No. 29, in press.

62. Hall, D. S., Haslag, K. P., Neff, S. G. 1975, unpublished.

63. Hall, D. S., Montle, R. E. 1975, unpublished.

64. Hall, D. S., Montle, R. E., Atkins, H. L. 1975, Acta Astr. **25**, 125.

65. Heard, J. F., Bakos, G. A. 1968, J. R. Astr. Soc. Canada **62**, 67.

66. Herbst, W. 1973, Astr. Astrophys. **26**, 137.

67. Hiltner, W. A. 1947, Ap. J. **106**, 481.

68. Hiltner, W. A. 1953, Ap. J. **118**, 262.

69. Hinderer, F. 1957, A.N. **284**, 1.

70. Imbert, M. 1971, Astr. Astrophys. **12**, 155.

71. Koch, R. H. 1970, I.A.U. Colloquium No. 6, p. 75.

72. Koch, R. H. 1974, A.J. **79**, 34.

73. Koch, R. H., Plavec, M., Wood, F. B. 1970, Pub. Univ. Pennsylvania Astr. Ser. 10.

74. Koch, R. H., Sobieski, S., Wood, F. B. 1963, Pub. Univ. Pennsylvania Astr. Ser. 9.

75. Kopal, Z. 1959, Close Binary Systems (New York: John Wiley and Sons).

76. Kraft, R. P. 1967, P.A.S.P. 79, 395.

77. Kriz, S. 1965, B.A.C. 16, 306.

78. Kron, G. E. 1947, P.A.S.P. 59, 261.

79. Kron, G. E. 1950, P.A.S.P. 62, 141.

80. Krzeminski, W. 1969, in Low Luminosity Stars, ed. S. S. Kumar (New York: Gordon and Breach) p. 57.

81. Kuhi, L. V. 1964, P.A.S.P. 76, 430.

82. Leung, K. C., Jurkevich, I. 1969, B.A.A.S. 1, 251.

83. Lloyd Evans, D. H. H. 1973, I.A.U. Symposium No.51, p. 140.

84. Lohsen, E. 1974, Astr. Astrophys. 36, 459.

85. Lucy, L. B. 1975, preprint.

86. Martins, D. H. 1975, P.A.S.P. 87, 163.

87. Mergentaler, J. 1950, Contr. Wroclaw Astr. Obs. No. 4.

88. Milone, E. F. 1969, in Non-Periodic Phenomena in Variable Stars, ed. L. Detre (Budapest: Academic Press) p. 457.

89. Milone, E. F. 1975, private communication.

90. Milone, E. F. 1976, I.A.U. Colloquium No. 29, in press.

91. Miner, E. D. 1966, Ap.J. 144, 1101.

92. Moffett, T. J., Bopp, B. W. 1971, Ap.J. 168, L117.

93. Montle, R. E. 1973, M.A. Thesis, Vanderbilt University, Nashville, Tennessee, U.S.A.

94. Mullen, D. J. 1974, Ap.J. 192, 149.

95. Naftilan, S. A. 1975, P.A.S.P. 87, 321.

96. Nather, R. E., Harwood, J. 1972, I.A.U. Circ. No. 2434.

97. Nelson, B. 1976, I.A.U. Symposium No. 73, in press.

98. Northcott, R. J., Bakos, G. A. 1967, A.J. 72, 89.

99. O'Connell, D. J. K. 1951, Pub. Riverview Coll. Obs. 2, 85.

100. Oliver, J. P. 1971, B.A.A.S. 3, 14.

101. Oliver, J. P. 1973, I.A.U. Symposium No. 51, 279.

102. Oliver, J. P. 1974a, Ph.D. Thesis, University of California, Los Angeles, California, U.S.A.

103. Oliver, J. P. 1974b, B.A.A.S. 6, 37.

104. Oliver, J. P. 1975, private communication.

105. Plavec, M., Grygar, J. 1965, Kleine Veröff. Remeis Sternw. Bamberg 4, No. 40, p. 213.

106. Popper, D. M. 1956a, Ap.J. 124, 196.

107. Popper, D. M. 1956b, P.A.S.P. 68, 131.

108. Popper, D. M. 1957, J. R. Astr. Soc. Canada 51, 57.

109. Popper, D. M. 1961, Ap. J. 133, 148.

110. Popper, D. M. 1962, P.A.S.P. 74, 129.

111. Popper, D. M. 1967, A. Rev. Astr. Astrophys. 5, 85.

112. Popper, D. M. 1969, B.A.A.S. 1, 257.

113. Popper, D. M. 1970, I.A.U. Colloquium No. 6, p. 13.

114. Popper, D. M. 1974, B.A.A.S. 6, 245.

115. Rigterink, P. V. 1972, A.J. 77, 230.

116. Rucinski, S. M. 1974, Acta Astr. 24, 119.

117. Sahade, J., Cesco, C. U. 1944, Ap.J. 100, 374.

118. Stebbins, J. 1928, Pub. Washburn Obs. 15, 29.

119. Struve, O. 1946, Ann. d'Astrophys. 9, 1.

120. Struve, O. 1948, Ap.J. 108, 155.

121. Struve, O., Horak, H. G. 1950, Ap.J. 112, 178.

122. Torres, C. A. O., Ferraz Mello, S. 1973, Astr. Astrophys. 27, 231.

123. Ulrich, R. K. 1976, I.A.U. Symposium No. 73, in press.

124. van't Veer, F. 1973, Astr. Astrophys. 26, 357.

125. van Woerden, H. 1957, Ann. Sterrew. Leiden 21, 3.

126. Vogt, S. S. 1975, Ap.J. 199, 418.

127. Weiler, E. J. 1974, P.A.S.P. 86, 56.

128. Weiler, E. J. 1975a, B.A.A.S. 7, 267.

129. Weiler, E. J. 1975b, I.A.U. Inf. Bull. Var. Stars. No. 1014.

130. Wielen, R. 1974, in Highlights of Astronomy 3, ed. G. Contopoulos (Dordrecht: D. Reidel Pub. Co.) p. 395.

131. Williamon, R. M. 1974, P.A.S.P. 86, 924.

132. Wilson, O. C., Wooley, R. 1970, M.N.R.A.S. 148, 463.

133. Wood, F. B. 1946, Contr. Princeton Univ. Obs. No. 21, 10.

134. Worden, S. P. 1974, P.A.S.P. 86, 595.

135. Wright, K. O. 1954, Ap.J. 119, 471.

136. Young, A., Nelson, B. 1972, Ap.J. 173, 653.

Discussion to the paper of HALL

SEGGEWISS: You compared the spot activity of the RS CVn stars with that of the sun. I see at least two major differences between the solar spot activity and the RS CVn activity. (1) The light variations of the sun due to spots is on the order of a few thousandths of a magnitude. You need several tenths of a magnitude (i.e., a tremendous activity) for RS CVn stars. (2) On the active sun we find a concentration of spots at certain latitudes, a latitude distribution. For RS CVn stars, you must assume a stable concentration of spots at a certain longitude interval.

HALL: Concerning the first point, you are entirely correct. In fact if you look at the KO IV component of RS CVn as an example, the brightness of the brighter and the fainter hemispheres are in the ratio 3:2. This must indeed represent tremendous activity. I mention that recently Mullen, in several articles in the Astrophysical Journal, has explored the possibility that the dynamo mechanism, which might be producing spots in our sun, can be scaled up to produce the large-scale spot activity observed in the flare stars and in the RS CVn binaries.

Concerning the second point, let me stress that there is one important difference between the sun and the KO IV star in RS CVn: the sun is a single star. It is my feeling that the presence of the other star in RS CVn, which is nearby and equally massive, is the basic reason for the dramatic longitudinal asymmetry. I cannot, however, even begin to suggest a specific mechanism.

A. WEHLAU: Do the unusually shallow secondary minima of RS CVn show a 10 year periodic variation as would be expected on the basis of the mechanism you propose?

HALL: If my model for RS CVn is correct, the depth anomaly in secondary minimum <u>should</u> show a ~10-year periodic variation, but no one has yet looked into this. The problem is that only one binary - RS CVn itself - has observational material suitable for such a test. The astronomers at Catania have observed RS CVn each year for about 12 years now, but as of now only 4 years of observations have been published.

GEYER: To comment on Dr. Seggewiss' remarks, I would like to mention that on the sun Wolf in Zurich in 1890 observed that sunspots for several years appeared always at the same solar longitude.

KWEE: According to one of your tables, most of your groups are detached systems. Yet you try to explain the period variations by variations in mass. What mechanism do you propose?

HALL: The basic idea is that there is strong mass outflow preferentially from the fainter hemisphere, induced presumably by the flare activity expected to be associated with spot activity. When the spotted region is on the leading hemisphere, the period should change in one direction; when the spotted region is on the trailing hemisphere, the period should change in the other direction. Ejection velocities associated with flare activity should be $\gtrsim$ 1000 km/sec, so there is no problem getting the mass to escape from these detached binaries. I think this mechanism is on the right track, but there are still problems and details to be worked out.

MATTEI: The properties that you gave for RS CVn variables are very much like those of T Tauri stars. However, you do not put the T Tauri stars with those like RS CVn variables. May I ask why? Few T Tauri stars are known to be spectroscopic binaries.

HALL: I did not include the T Tauri stars in my list of similar systems because they tend to be single stars. But the point you make is an important one. In fact, if you looked at the KO IV component of RS CVn alone, it would be difficult to distinguish from a T Tauri star in its physical characteristics (mass, radius, luminosity, temperature) and various peculiarities (H and K emission, Hα emission, light variations, ultraviolet excess, infrared excess, and mass outflow).

VAN HORN: I have two questions for you. First, if your mechanism is correct, there must be magnetic fields in these stars. Do you have any idea how large these fields are? Second, with respect to Dr. Lomb's question, do you have some idea of how large the mass-loss rate may be?

HALL: First, to my knowledge no one has ever looked for magnetic fields in these stars. This is an important observation which indeed should be made. I have been suggesting this for several years, with no apparent result, and I can only suggest it again at this time.

Second, I can say that if you use a simple impulse-type mass loss mechanism and assume ejection velocities of 3000 km/sec, you require 10^{-6} $M_{\odot}$/yr to account for the observed period changes in RS CVn. This is very large, certainly too large. But Dr. Peter Biermann has suggested to me that the ejected mass probably co-rotates with the system out to some Alfven radius which is many times larger than the semi-major axis of the orbit. If so, then the required rate of mass loss can easily be reduced substantially.

BUDDING: How sure can we be that the period changes referred to are real and not just the photometric displacement of minima by the "wave" variation? For example, Binnendijk referred to O-C variations in the minima of YY Gem, but later Kron discounted these as being probably effects of the out-of-eclipse light distortions on the precise positions of the minima.

HALL: This is a very important question to raise. The light curve asymmetry can and does distort the shape of primary eclipse and does generate O-C changes which are spurious in the sense that they do not represent real changes in the orbital period. It turns out, however, that in the majority of cases the observed O-C changes are very many times larger than could be produced by such distortion. The clearest example is SS Cam. Here the total range of the O-C variation is more than a half day, which is more than the angle of external contact for primary eclipse! This matter is discussed in detail in a paper which I have submitted to Acta Astronomica. But I want to join you in emphasizing that the spurious component in the O-C variation of every binary must be recognized, treated critically, and removed before any statement is made about true period changes.